The Balance

THE BALANCE
A FIREFIGHTER'S GUIDE TO LIFE IN THE SERVICE

JOHN AMBROSE

FIRE ENGINEERING • BOOKS •

Disclaimer

The recommendations, advice, descriptions, and methods in this book are presented solely for educational purposes. Photos are for instructional purposes only. Always wear the proper level of approved personal protective equipment when conducting training drills and operating at incidents. The author and publisher assume no liability whatsoever for any loss or damage that results from the use of any of the material in this book. Use of the material in this book is solely at the risk of the user.

Copyright © 2026 by
Fire Engineering Books
110 S. Hartford Ave., Suite 200
Tulsa, Oklahoma 74120 USA

800.752.9764
+1.918.831.9421
info@fireengineeringbooks.com
www.FireEngineeringBooks.com

Executive Vice President: Eric Schlett
Vice President, Group Publishing: Amanda Champion
Vice President of Content Operations: Starlet Franz
Sales and Customer Service Manager: Lane Nash
Managing Editors: Diane Rothschild and David Rhodes
Production Manager: Tony Quinn
Senior Development Editor: Daniel Edward Petrino
Book Designer: Robert Kern, TIPS Publishing Services, Carrboro, NC
Cover Designer: Brandon Ash

Library of Congress Cataloging-in-Publication Data Available on Request

ISBN: 9781593705206

All rights reserved. No part of this book may be reproduced, stored in a retrieval system, or transcribed in any form or by any means, electronic or mechanical, including photocopying and recording, without the prior written permission of the publisher.

Printed in the United States of America

1 2 3 4 5   30 29 28 27 26

*I would like to dedicate this book to the
three powers of the fire service:
Those who came before us—
they blazed the trail for us to follow.
Those who are with us—
their courage keeps the dignity of our endeavor alive each day.
Those who have yet to come—
they will carry on a legacy of bravery that
will always keep us on a pedestal of reverence.*

Contents

Foreword xi
Preface xiii
Acknowledgments xv
Introduction xvii

1 What Is the Balance? 1
   A Convergence of Stories 1
   What Are We Striving for? 3
   Reflective Prompts 6

2 The Firefighter: An Unusual Creature 7
   Grace 7
   How We're Wired 9
   Choices 11
   Final Thoughts 12
   Reflective Prompts 13

3 The Roles We Play 15
   How Legends Operate 15
   Options 18
   Forks in the Road 19
   Roles 20
   Reflective Prompts 24

4 The New Hire: The Future 25
   Ready or Not 25
   We've All Been There 29
   A Smooth Transition Is Key 32
   Final Thoughts 32
   Reflective Prompts 34

5   Midlevel Firefighters: The Present 35
    A Perfect Storm 35
    Teamwork Matters 40
    The Forgotten Tool 43
    Final Thoughts 44
    Reflective Prompts 46

6   Senior Firefighters: The Past 47
    A Peaceful, Easy Feeling 47
    Go Time 49
    The Relevant Past 52
    Change 53
    Stay Fluid 54
    Keep the Play Pen Intact 55
    Final Thoughts 56
    Reflective Prompts 58

7   The Lieutenants: A Conduit Like No Other 59
    Tough Decisions 59
    Two-Way Traffic 65
    It's Just Different 65
    The Jack of the Trade 66
    Reflective Prompts 70

8   The Captains: It's Your Shift—Sort of 71
    Out of His Element 71
    A Different Kind of Transition 76
    A Culture of Your Own 77
    Play the Game 79
    You Reap What You Sow 80
    Final Thoughts 81
    Reflective Prompts 82

9   The Chiefs: Where Culture Is Dictated 85
    Biting Tongues 85
    Tools of the Trade 88
    The Balancing Act 89
    Transparency Equals Success 91
    Roots 93

Contents ix

    Final Thoughts 94
    Reflective Prompts 95

10 Two Homes. 97
    Oops 97
    Manifestations. 102
    Finding the Balance Between 104
    Your Houses, Your Homes 107
    Reflective Prompts 107

11 Dealing in a Healthy Way. 109
    A Close Call 109
    On Our Own Path. 114
    An Easy Cure 116
    Endnotes 120
    Reflective Prompts 120

12 All Shapes and Sizes. 121
    A Fish out of Water. 121
    Different Strokes. 124
    The Big Three. 125
    It's Your Time! 132
    Reflective Prompts 132

13 That Darn Ego 133
    The Eyes of the Devil. 133
    The Rolling Eyes of the Watercooler 136
    The Two Victims of the Ego. 137
    Breaking the Cycle 139
    Three's Company 141
    Reflective Prompts 141
    Note. 141

14 Empowering and Its Benefits 143
    Out of Order 143
    The Power of Empowering 147
    Empowering Those Who You Manage. 148
    Empowering Those Who Work with You. 149
    Don't Underestimate Your Team 151
    Reflective Prompts 151

15 All Together 153
- The Three Traits That Make Us . . . Us 153
- The Mind. 154
- The Body. 156
- The Spirit 158
- It's a Wrap. 160
- Reflective Prompts 160

Index 163

Foreword

I have known John for over 20 years. We were classmates in paramedic school before being hired just 6 months apart at Avon (OH) Fire Department. We became friends quickly as we bonded over a love for movies and our common agony of being Cleveland sports fans. Over the past two decades, we have shared great memories and helped each other through rough patches in our lives. He's truly one of my best friends.

As a firefighter, John was able to win people over with his personality and humor, but his laid-back demeanor was at times difficult for some to interpret. John had skeptics within the department who had a different perspective of the job. As he grew into a senior member and eventually a company officer, members began to buy into his approach to leading. His respect among the members and influence within the department grew as his career did. Since his promotion, those once skeptical of him praised his ability to lead his shift with sound decisions. Quite honestly, he is a rare example of a good firefighter who became a great officer. He has been an extremely welcomed addition to our officer corps.

John has been able to achieve such success by indeed finding the balance in the fire service. We are tasked as officers with both managing and leading. The fire service has told our leaders that you must be a general, a mentor, a teacher, a marriage counselor, a coach, a therapist, and a friend—with no training. John has handled all these tasks with great results.

We ask our members to juggle the fireground with in-house issues or offset the emergency scenes with personnel disputes, while being provided limited references for any of these challenges. Consider this book that long-awaited reference. Hopefully, this book will help you find your balance as you navigate the often-uneven terrain of the fire service.

I hope you find this book a useful tool as you continue your journey through the best job in the world. And don't forget to have some fun!

—James Fischer
Assistant Chief, Avon Fire Department

Preface

The Balance: A Firefighter's Guide to Life in the Service is an informal guidebook for firefighters of all levels to achieve success in their pursuit of a fruitful career. Tips throughout will help the reader with common concerns in our everyday lives in the service. My intentions are to inform but not overwhelm, to engage but not judge, and to challenge but not patronize.

Many of the concepts described in this book are hopefully being implemented in your career already. The chapters contain suggestions on how to handle many problems we all face from time to time on an interpersonal level. Although this book has sporadic tactical content, most of the text focuses on how we carry ourselves on a daily basis in how we interact with others.

Much of what I have written is based on personal observations and applications to solve problems I or someone I've known have faced. Throughout the book, fictitious characters face tough calls and pivotal moments that determine success or failure in their scenarios. Hopefully, their brief journeys bring to light something you can relate to in one form or another. Although these characters and their challenges may not align perfectly with real-life situations, they may still strike a nerve that translates to a familiar problem you may have had.

My analysis of these scenarios combines personal opinion and sound thinking. However, the advice and recommendations are not to be taken as gospel. On the contrary, they should spark conversations and successful approaches that will hopefully lead to good solutions to common problems.

Ideally, this book can be a blueprint for success as members of your department work their way up the ladder of experience and rank. Each level, from new hire to chief, is addressed in easy terms. Along the way, there are chapters that address all of us together as well.

The Balance was devised to keep my fellow firefighters on a course of success, with rewarding results. Certain elements will require self-analysis. Successful readers of this book will draw on their own ability to self-reflect and be open to change. Ultimately, this book is designed to open readers up to the possibilities we all have within ourselves. Whether you take the necessary steps to better yourself remains purely up to you.

Enjoy your time reading *The Balance*. Remember, we are all in this together.

Acknowledgments

I am grateful to the many people who inspired me to write this book. Thanks to my former captain, Rodney Meadows; he taught me that consistency and work ethic go a long way to respect and productivity. Thanks to Assistant Chief James Fischer, who has been my best friend on the Avon (OH) Fire Department for 20 years; every day he reminds me of the possibilities this job offers and how determination and hard work translate to results that carry over to every member of our department.

Thanks to my generational inspirations—my parents and my kids. Dad, you were always the dreamer who taught me to look beyond the bounds of the ordinary. Mom, you have personified love and loyalty at every turn of my life. My son, Jackson, has an unwavering moral compass that would inspire people twice his age. My daughter, Bryn, shows me her fierce independence from a place that remains a mystery to this day. Love you all, dearly.

Finally, I would like to pause and remember the 343 firefighters who were lost on 9/11. On that fateful day I decided to become a firefighter, and my small contribution to the fire service will never do justice to the ultimate sacrifice they all made. These lost heroes, along with the other first responders who perished, are and will always remain an inspiration to me.

Introduction

The Balance: A Firefighter's Guide to Life in the Service takes a simple approach to common obstacles we face in the fire service. My goal in writing the book was to replace an inundation of recommendations with a few basic guidelines that are easy to understand, thoughtful in their approach, and applicable not just at the firehouse.

The fire service was never a path I thought I would choose growing up. As kids, we dream of what we may become one day. We imagine ourselves in different scenarios and as the hero of these stories crafted in our heads. To be honest, playing the part of a hero running into a burning building never entered into my childhood fantasies. Even as I grew older and became a young man, I wasn't quite sure where I was headed with my life.

When 9/11 took place and shocked us all, I refocused my priorities: I dropped out of the business world in the private sector and decided to do my part by becoming a firefighter. I enrolled in the fire academy and went to paramedic school at night. There were times when I wasn't sure exactly what I was getting into. But I stuck with it.

After numerous tests and interviews, I became a full-time firefighter for the city of Avon, Ohio in May 2006. Since embarking on this adventure, I have enjoyed a long and fulfilling journey. I've made plenty of mistakes and tried my best to grow from them as I've rolled along with my career. Being a part of the sacred fellowship of firefighting has allowed me to reflect on my life in the service and what I've learned along the way. Now in my 19th year in the fire service, I have observed a great many events and situations that have prompted me to pause and reflect on what it is to be a firefighter, how to rise through the ranks, and how to forge terrific relationships with my brothers and sisters in the service.

Writing a book was never on my radar until it dawned on me that the reflective qualities I have developed would have a positive effect on those traveling the same path as I had in this industry. Over time, I've been able to hone certain skills that—when applied correctly—work. This book tackles fundamental challenges that we face in our quest for happiness in the profession we love.

Most of us are dedicated to this wonderful life of sacrifice and service in an effort to feel fulfilled and content as we grow older.

My goal is that this guidebook will provide methods that aid firefighters at every level in their own journey toward success. Some of these may seem like common sense. Others may offer a fresh perspective on old problems. Most chapters start with fictional stories followed by analysis and questions to prompt self-reflection. Like any book of this nature, the more you put into it, the more you will get out of it.

The pages that follow are written in frank and honest terms. I refuse to get hung up on technicalities and therefore have written a book that addresses universal issues. Our job isn't all about fighting fires and saving lives. On the contrary, this book frequently addresses the 90% of the time during which we are paid to be ready and coexist with others paid to do the same. Our well-being should always be at the forefront of our lives. Too often, this crucial element is lost, and we can all use a little tune-up to get us back on track and remind us how fortunate we truly are.

I'm not proclaiming myself an authority on any particular topic in the following chapters. I'd rather the reader consider me a resource, someone who can relate to what they go through as a fellow member of the fire service. You may not agree with everything you read, and I admire that instinct—which tells me that the wheels are turning, you're thinking things through, and you are engaging with the words on the page.

Now I'd like to invite you to sit back, read on, and open yourself up to introspection. You may pull something out of these chapters that was never intended to have the effect it does on you. I applaud you for taking an individual and proactive approach to self-analysis. I feel you are well worth your time and attention. I hope you agree. Thanks again.

1 What Is the Balance?

There is no such thing as work–life balance—it is all life.
The balance has to be within you.

—Sadhguru

A Convergence of Stories

A fresh new cadet named Max sat in the back of Ladder 46 as part of a commercial fire alarm for a possible structure fire on the west side of town. They were staging as a support crew for the first-responding engine that had arrived on scene 2 minutes earlier. The crew were thrown together because of two call-offs right before shift. He glanced toward the truck operator and the officer as they sat in the front seats waiting for the 360° evaluation to be completed by the first-arriving officer.

This is the path. One day I'll be where I'm supposed to be, Max told himself.

Whenever the fire academy or paramedic clinical rotations began to take their toll, that's what Max would think. But here he was. He had arrived. At 21 years old, his long-term savings account was up and running—and he had a full bank account compared to his friends from high school. He didn't have a steady girlfriend, didn't really have any bills yet, and had purchased the pickup truck that seemed to be standard issue for the personnel at the firehouse. He was convinced he would be a millionaire once he completed the drop in his late 50s. Good things were going to happen to Max, and he could feel it finally paying off for him as he gazed at a yellow-tinted streetlight across the street from where Ladder 46 was parked.

Max thought about his parents and how they had become snowbirds now that his dad had retired from the bank. They bought a condominium in Florida and spent 4 months of every year soaking up the sun. Max's brother, Ryan, was in real estate and made decent money. He lived about 2 hours from Max, and they got together when they could. Melanie, Max's oldest sibling, was married and estranged from the family. She and her husband rarely came around, and a phone call from her even monthly to her parents was not typical. A rift a few holidays back had fractured the family when an argument had turned into a fight. Melanie dug her heels in and so did Max's father. Time was slipping away, and this had cast a shadow over an otherwise exciting time in the young firefighter's life.

At that very instant, Sam glanced at the same yellow streetlight from the truck operator's seat. He had been on the department for the better part of 9 years and was desperately trying to hold on to the passion he once had for the job. He was on the executive board of the union, and negotiations were right around the corner. Another round of nonsense from a city that had piles of money in reserve yet couldn't cough up a rusty nickel for the fire department when it came to simple raises to keep up with the cost of living. The 33-year-old had always tried to look on the bright side during his first few years on the department. But he could feel a change happening.

Some of the senior members were beginning to wear on Sam with their negativity, and he was afraid he was heading down that same road. And why not? Behind closed doors, it was much easier to hurl rocks than deflect them.

Sam had two young children at home: Ella was 6 and Garrett only 3 years old. His wife, Denise, was once his moon and stars; now she felt like a business partner and a roommate as they tried to navigate work, finances, family, activities, and anything else that came their way on what seemed like a daily basis. Sam still loved his wife but in a way that wasn't quite as magical as it was when they tied the knot 8 years prior.

It'll get back to where it was, Sam would tell himself. It had to, right?

Sam's mom was still healthy, but his father was beginning to show the signs of wear that years of drinking Johnnie Walker usually brings to the party. Cirrhosis was the preliminary diagnosis, and Sam had been a medic long enough to know that a rocky road lay ahead unless drastic changes were implemented sooner rather than later. That was easier said than done with his dad. Ah, the constant struggle between being a steady force of reason and an enabler. He turned and shifted his weight, briefly looking toward the front door of the firehouse that still had nothing showing from the Alpha side.

Greg, the acting officer in charge for the shift, was gazing at the same front door. He was on year 1 in the DROP retirement program and on year 26 in the field overall. He was solid by all accounts in the ranks. He had been promoted to lieutenant about 7 years prior but lost by a whisker on the most recent promotional examination for a vacated shift captain's spot. He was now 50 years old, and some days he really felt like it. His joints didn't move the way they used to, and he was finding it harder to catch a full night's sleep, even during a string of vacation days.

Both Greg's kids were in college, and he worked part-time for his brother's contracting business on the side to occupy his time. He had been divorced from Kelly for almost 5 years, and it took 3 of those years to find himself again. They had been married for 24 years and steadily grew apart. His daughters, Chloe and Jane, were doing fine in school and, after a rough first year, had adapted to living with split parents. Kelly and he were amicable, which helped immensely as they tried to navigate starting over in their 40s without each other.

Greg was a drinker. Perhaps not to the extent where an intervention was needed, but definitely to the point that even he felt as though he should slow down. He liked to think it didn't contribute to his divorce; still, people change when they drink, and he was no exception. He never came to work drunk, and he never came to work stinking like a bad hangover. There had been a line of drinkers in his family, and he knew this family history. Indeed, an addictive streak ran through his family. His older sister Carol had a nasty relationship with Percocet for years to deal with "back issues" that never seemed to go away. His younger brother Doug had three citations for driving under the influence on his record and never could get himself together long enough to sustain a decent job.

The fire service grounded Greg. He felt accountability to his department and to himself to show up and do his job the best he could. He earned recognition the day after his 34th birthday for leading a rapid intervention crew that rescued a downed firefighter during a collapse at a split-level residential fire. The firefighter survived but went on disability because of damage to his right leg and ankle that never healed well enough to allow his return. Greg did his job that day and received the admiration of the young members and respect from the senior ones because of it.

All three firefighters sat in an undermanned apparatus doing what they were paid to do. They each had stories reflecting different backgrounds and different circumstances. Many of us can relate to some degree or another with all three of the firefighters we just met. As we work through all of the problems that arise in the field we've chosen, there is one condition we all strive for to make sense of our purpose both at work and at home: balance.

What Are We Striving for?

Balance has many different meanings in a mechanical sense but speaks volumes when applied figuratively to life situations. Scores of advice and sayings have thrust this word into mainstream jargon and plastered it across self-help posters. But if we truly want to identify what it is that creates the wholeness and feeling of well-being we all crave, then we must accept balance as the central component we need to strive for. It is an equilibrium we find in our lives on a personal, professional, and emotional level. When the scales tip drastically to one side or the other, we are left vulnerable and unprepared for the uncertainties our job presents. Regardless of whether we recognize it, in every situation we encounter or problem we solve, a balance needs to be obtained to achieve the ideal (or sometimes simply acceptable) results. The risk-versus-reward dilemma applies to each decision we make.

A couple who have been married for a decade may not speak to each other for days on end. Is that normal and comfortable for them? Perhaps. Is it healthy and fulfilling for them and their marriage? That is a different matter. Comfort does not always equal balance. You will most assuredly feel comfortable and relaxed when the balance is achieved. But fooling yourself into believing you have achieved it is disingenuous to you. And that line in the dirt gets smudged when we settle into routines.

Ideally, this book serves as a guide to navigate the very rewarding yet challenging life we've chosen. As firefighters, there is a nobility that accompanies us no matter what we do. There is also a reason we are looked upon with admiration and respect. We have chosen to do our jobs regardless of the outcome. In any instance, regardless of the circumstances, the people we have sworn to protect have called us in their time of greatest need. Whether we are walking a parade route, working a full arrest with family crying nearby, or pulling a cross lay for offensive operations attacking a room and contents fire, we must be able to do our job efficiently and with unwavering determination and coordination.

Countless books have been written on fire suppression tactics. Many others have offered operational advice for when we go into battle with horrible conditions and situations given our tools and resources. But this book is designed primarily to help us in what sometimes feels like the most confined space in our lives: the space between our ears. Breaking it down into pieces will advance our journey together. As we move forward, remember that we have all heard the clichés regarding fellowship and family in our stations across the country and even on a global level. So often our commitment to the job blinds us to our critical commitment to ourselves and each other.

As with so many people in society as a whole, we have decided to look outward to help us satisfy the needs we have. As the great spiritual teacher Sadhguru states, the balance must be within you. Therein lies the key. The place we see when looking inward is the one true place that can never be taken from us. We do this through self-reflection. The little voice inside our heads (we all have one) is also watching, always observing. Listen once in a while. That voice is smarter than you give it credit for, and it's been with you since the day you were born.

The balance is what gives us a feeling of control over ourselves. No matter what we do, we can never control anyone else fully. However, if we can start practicing the self-care we deserve, a wonderful transformation away from negativity and pessimism begins to take hold of our attitudes, our relationships, and our health. When the balance is reached, we can move forward toward goals in a positive and fruitful way with meaning and initiative. It's never too late to dive in. Before you know it, doors open, relationships with the people at the station blossom, morale soars, and the home front settles.

We will always face challenges in our careers and in our personal lives. Stress rears its ugly head at the worst times, it seems. Yet, when we have donned the armor of protection through the work-life balance we learn and practice, stress evaporates faster than a fog stream at a ceiling ready to flash over.

This book primarily deals with the fire service because that's where I've been for the past 19 years. Nevertheless, many of the reflective components are also applicable to other careers and facets of the busy lives people lead today. We all need to work, pay bills, juggle relatives, go shopping, cart kids around, take care of households, cultivate friendships—the list is endless. Having a solid foundation helps when we must wade into the deep end of responsibility. Hopefully this text can be a helpful tool in your repertoire that assists you in your navigation along an exciting but often daunting path.

I have included a set of reflective prompts at the end of each chapter to assist you. Use whatever looks interesting to you. Hopefully, something will strike a nerve with you somewhere you never thought it would. If you feel you have everything figured out and don't need to improve on anything in finding your balance, you are *way* ahead of the game. If not, it's OK. I have a long way to go myself.

As in this chapter, you will be introduced to different characters who have been placed in situations that test their ability to find the balance. The three characters you met at the beginning of this chapter were not subjected to situations where they were forced to make life-or-death decisions. Yet, after reading about them, you got a glimpse into what each individual was dealing with in their personal lives. Remember, there is a balance to be found in every task you are given, every decision you make, and every relationship you enter.

How will you know when you've found the balance? You have already found it in many parts of your life. Look inward and the answer will be there for you. You'll realize you've found the balance when you feel inner peace and tranquility when you think of that situation. A calmness creeps in when you have made the right choice, are confident in your decision moving forward, and have a springboard to move on to the next challenge life throws your way.

There are so many possible examples of being lulled to sleep by routine. We must take stock in those routines and ask ourselves if we are settling with what is or striving for what could be. In every situation, we have choices to make when given a fork in the road. In the short term, it could be as simple as a menial task like taking out the garbage. On a grander scale, it could dictate our self-image and the respect of all the people in our lives—it's that important. As you read the chapters that follow, take stock of your own life and all the decisions you make within it. You'll be amazed how many times a day you must find the balance.

Reflective Prompts

1. Our minds race when we are in "hurry up and wait" mode while we are on a call, waiting to spring into action. How are you able to clear the noise in your head and get ready for the tasks you are about to perform?
2. If there are distractions running through your mind, what are they about? Work? Home? How are you able to put those thoughts on the shelf when it's go time? If you can't pause those thoughts, have you looked into strategies to help?
3. Do you consider the station a part of the balance in your life? Do you use home or the firehouse as an outlet to escape from other parts of your life that you feel are lacking?
4. Take a mental note of what you think about on your way to the station next time. Is there a pattern to your thoughts, or do you drift depending on the day and what it presents?

2 The Firefighter: An Unusual Creature

When a man becomes a fireman his greatest act of bravery has been accomplished. What he does after that is all in the line of work.

—Edward Croker
Chief, New York City Fire Department (FDNY)

Grace

Darius sat at the festive table with his wife, Kim. They had arrived a little late that Thanksgiving after the typical logistical chaos of a young family during the holidays. He didn't mind. Family chaos was better than station chaos any day for him.

This was Darius's first Thanksgiving off in 2 years, and it was nice to take the break. He had waited his turn in line for time-off picks and finally could take a major holiday off. He'd been on the department for 11 years and was ready to take some time for himself. He and Kim had been married for 9 years, and their children, D. J. and Kendra, were awesome.

The young family sat at the dinner table with Darius's parents, several extended cousins, two aunts who usually fell asleep by the time dessert hit the table, kids of different annoying ages, and his older brother Andre. Andre was 3 years older than Darius, and although they grew up very close, as they got older, their relationship began to splinter. Nobody would point the finger at Andre's overbearing wife, Rochelle, but some things don't need to be proclaimed from mountaintops to be evident.

There was a pattern to holiday get-togethers over the years. The food was delicious, the kids would buzz to and from the table with questions and problems while running way too fast through the house, and Andre would get drunk and begin to complain about how stressful his life was at the insurance company he worked for. His wife would add some extra problems for good measure. The table talk would be dominated by unfair commission checks, skyrocketing premiums, and the unbearable sob story of how Andre and Rochelle found it impossible to survive on $200,000 a year.

Kim was a sport about it. She loved Darius and had been his wife for over 9 years. Their two kids also got along well with Andre and Rochelle's three girls. But each year the unfounded claims of life's cruelty took its toll on poor Kim. She silenced her objections pretty well. This year was no exception. Five vodka sodas in and Andre was up and running.

"So then Furgeson actually has the audacity to tell me that there still isn't enough new business coming in," he said.

Rochelle added, "Like he doesn't bring in enough money for his boss as it is!"

"Well, honey, I'm sure you're doing a great job," Darius's mother replied as she took a nibble of pie.

"It seems unbearable sometimes. I don't need this stress!"

"I'm sure your hard work will pay off," Darius's mother answered.

The complaining lasted through dessert and soured any possibility of board games afterward. Darius listened to his brother, tried to steer the conversation to lighter topics, and with a glance from his wife, knew it was time to head home.

They lived a 45-minute drive from his parents, and both kids dozed off in the backseat. Kim decompressed as she looked out the window, and Darius could tell she was shaking off her usual post-Thanksgiving frustrations.

"Thanks, babe," Darius said as he took her hand lightly.

"I just don't get it," she said as she allowed it. "Every year we listen to the same garbage, and every year they have no idea what you go through at work on a daily basis."

Darius glanced at her, and a small grin crept across his face. Kim caught the glance and playfully giggled as well.

"No, seriously! You had that lady that burned up last week at that fire and that really bad car crash the week before!" The smile fell from both their faces. "I wish once you'd put them in their place and give them a dose of what real stress at work is like."

"You know that's not gonna happen," Darius said softly. "Not my style."

"But couldn't you try? It seems to stress your mom out so much."

"There really is no point. He's been that way his entire life," Darius said. "Besides, the only one I really open up to is you anyways. The last thing I need is his opinion about handling real problems at work when he can't even handle fake ones."

Kim grabbed his hand and kissed it before looking out the window as Darius cranked Marvin Gaye up a couple ticks on the radio. She knew he was right. Talking about that stuff was never his style.

How We're Wired

Darius presents a great example of what a lot of us do in the service—simply put, we do our jobs. Firefighters are cut from a different cloth, and we take pride in knowing that we have our game faces ready when we get to the station. We're always on the lookout for the next true emergency.

One great trait most of us have is the ability to compartmentalize potential emotional trauma sustained on calls. Importantly, when we take our oath to serve, we are bound to a professionalism few people can fathom, let alone handle in a healthy way.

"I could never do what you do," I hear on occasion when I tell people I'm a firefighter. Most of the time, such comments are prompted by preconceived notions of the bloody scenes we arrive on and of the dangerous conditions we encounter when we enter a burning building. Although these may be true to an extent, the average person sees exaggerated scenarios on television and film that depict us as fearless heroes who scale ledges for jumpers or streak through a fully involved house calling out for victims despite limited visibility. What they don't realize is the beating our emotions take when confronted with the reality of what we do.

When we don our personal protective equipment and enter hazard zones on the fireground or put on our gloves when trying to jump-start a dying heart, we become *process people*, instead of feeling ones. When feelings get in the way, things go wrong. We have been trained to do X in this situation and Y in that one. Worry about the feelings later—we'll get to those when it's time. Great theory; however, the problem is that we're human too. We try to suppress our vulnerability to keep the mystique alive for some and simply to survive for others. And that's where things become more of a challenge.

We've all heard the stories of bygone eras where self-contained breathing apparatus were considered to be only for wimps and how nobody nowadays knows just how good we have it. Maybe there's a lineage to your service; for example, you might be proudly following in the footsteps of a parent or sibling. There is definitely still an honorable circle we are a cohesive part of, and we have to realize that advancements in our equipment and execution are essential for our success. But death is still death. We are still responsible for our actions—or inactions—when faced with the decision to risk our lives for someone else's.

For us to reconcile the magnitude of what we do in our own heads, we have to be gentler with ourselves. Too many times, firefighters lose themselves because they become caught up in the persona they portray at the firehouse, and they continue with that portrayal in other areas of their lives. That's a slippery slope.

As firefighters, we are indeed a rare breed. We are warriors in a paramilitary organization. We are compassionate allies to a public that needs our help. We are fearless brothers and sisters who confront danger on a level few could conceive of. Those are the ideas great T-shirts are made of.

But we are also members of a miniature social club. We are jokesters. We are goofballs who love our sports and who cook large meals for our fellow firefighters who sit together around a large table and dispose of the swear jar we so carefully try to keep empty at home.

Although ours is a unique profession, nurses, doctors, police, emergency medical technicians, and even veterinarians share tasks similar to ours. All are called into high-stress situations and paid to perform on a level sufficient to preserve a life using split-second thinking, yet must be ready for devastating consequences after one false move. The tiny difference between ourselves in the fire service and our colleagues in other emergency areas is that we spend roughly a third of our careers actually living with our coworkers. Couple that with a severe dose of alpha personalities and other potentially annoying tendencies, and we get a supercharged environment of tension and resentment if we aren't careful. Aside from our law enforcement counterparts and emergency medical service crews, all other emergency responders work in a controlled environment conducive to success. I have yet to hear an emergency-room physician declare "scene safe" when a patient rolls into a trauma bay.

Another aspect that firefighters must deal with is the evolution of our roles within the ranks of the department we serve. As I will touch on in later chapters, our roles must incorporate fluidity as we move along in our fire service careers. Within any job, plateaus are reached and progressions to the next level are made. With that comes added responsibilities and privileges. But there are few other occupations that require a person to do as many things physically at 55 as a coworker can do at 25 years of age. If we can't find the balance within us and allocate our energy toward the pressing matters at hand, our approaches to our tasks begin to lose focus and we become disorganized in our execution.

We are unusual creatures. Whether this is for good or bad depends on the situation and how we handle it. As former New York City Fire Department Chief Edward Croker said, we have already made the heroic decision to join the service. After that, we are performing our duties in the line of work. Still, always remember why you chose to join in the first place. Stop and think. Evaluate the motivations that got you to the place you are and take the time to appreciate yourself.

As is human nature, we tend to move on to the next thing, to reach personal goals and benchmarks in our professional lives. What we fail to do, though, is stop and look around once in a while to take in exactly what it is we do for a living. People say that when we start in the fire service it isn't for the money. This is true in a monetary sense, but when we take stock of the richness of the jobs we perform and the tasks we are assigned to manage, we are left with a feeling of pride and accomplishment few others will ever enjoy.

Luckily, excellent resources such as employee assistance programs and outreach programs are available to get us over the bumps in the road. Once these resources are put in place, it becomes simply a matter of using them as intended. The ability to take that initiative rests squarely on the individual: each of us must make a choice that will help achieve the balance.

Choices

Unfortunately, sometimes we stray away from the best choices we have in dealing with situations. The pressure cooker of difficult situations on the scene, sleep deprivation, challenging citizens, and differences regarding station operations and policies all add stress to our job. How these unfold is not up to us; how we deal with them absolutely is.

Often, we turn to numbing agents to combat these obstacles, and this usually ends poorly. Female firefighters have an even tougher row to hoe because of historical industry bias and preconceived notions of inadequacy and physical limitation. Drinking, smoking, drugs, infidelity, food—we can go on and on with the vices to which people will bow down to salve the wounds we acquire at work. Having a plan in place helps. Having the balance absolutely does. (For more on these detrimental choices, see chapter 11.)

Often, we find ourselves dealing with the wreckage of poor decisions instead of finding the balance to absorb a blow to our psyche and our emotions. We are creatures of habit (I once told a member on my shift that we will make our beds about 2,800 times in a 25-year career. Let that sink in).

Not all bad choices take such an overt, dramatic tone. Subtle changes can occur as well, such as developing a short fuse with your spouse or children. Usually, this is because we haven't recognized the stress beating us up at work and we bring the stress home with us. If this sounds familiar, you're not making history. But you are making *your* history. Be aware of it. (We'll discuss that more in chapter 10.)

We are human beings asked to do tasks that would make many people on this planet cringe. Recognizing this helps us to process things and get our thoughts together before an incident takes place. Being proactive instead of reactive is much better than the other way around. The best way to see where we stand on the ladder of self-awareness is to take a good look at ourselves and realize that we are indeed unusual creatures.

Final Thoughts

The following points exemplify how unique we truly are in the fire service:

1. *We deal with death.* This is always at the forefront of the public's perception of us. The *what* has been established. Death is a natural progression for everyone. *How* people die is not established. We have a front-row seat to a horror show. Only through experience and developing coping mechanisms are we able to harness the resultant stress in all its magnitude.
2. *Our reputation precedes us.* One of the largest internal battles we face is with the reputation we have in the public's eyes. There is a push-pull relationship between two very distinct self-proclamations: On the one hand is our duty to earn the accolades and to honor the badge; on the other is the ego inflation that accompanies a distinguished title out of the gate. We need to honor the former and be aware of the latter. The two can coexist.
3. *The whole family must adapt.* Our entire family becomes accustomed to our lifestyle and our schedule. When I was first hired on full time, my spouse and kids had to adapt to my being gone every 3rd day. This can be taxing on everyone, especially with missing birthdays, holidays, and family activities.
4. *We are called upon as calming entities.* I wish I had a nickel for every time someone at a party said they were lucky I was there in case something went wrong. They acted as though I drove a tanker truck to the barbecue and had a booster line hooked up, ready to go. I would shrug and play along the best I could. Still, we are trained to be cool and collected in stressful situations away from the station, which is a great trait to have.

Unfortunately, we are anointed as the resident problem-solver just because of our job.
5. *Our coworkers depend on us.* Our line of work is unique. Many other occupations call on their peers to get matters done; paperwork needs to be filed, tasks must be completed, sales must be made, and so on. But few others put their lives in each other's hands. It's a pressure cooker of stress. Solid training and learning through experience allows us to prepare for these crises with good outcomes. There is no greater vote of confidence in someone than to put your life in their hands.

We indeed have a lot to live up to, and when we are sworn into the fire service, we adopt a role. Most of us are aware of this already. Nevertheless, sometimes we forget that we play roles all the time. The ultimate goal is to play only one role: ourselves. But that isn't always easy. The roles we play will be discussed in detail in the next chapter.

Reflective Prompts

1. How do your experiences and your lifestyle compare with those of other people you know? Do you consider your career and your experiences "normal"?
2. How has your family adapted to the life you've chosen? Have they settled in, or do they struggle? Do you struggle at times?
3. Are you confident in your abilities to answer the call if you need to save one of your coworkers? Are you confident in their being able to do the same?
4. In your experience, are you treated any different when people hear what you do for a living?

3 The Roles We Play

You can easily judge the character of a man by how he treats a person who can do nothing for him.

—Malcolm S. Forbes

How Legends Operate

Ray was "The Legend." On the fireground, he always seemed to be in the middle of the action, and his achievements were endless. He had been with the department for over 20 years and had seen enough fire to tell stories for hours on end. He worked in one of the busiest firehouses in a very large department. Many of the stories told about Ray were of impossible grabs, wild collapses, and dramatic bailouts. Through embellishment over time, the tales told about him grew taller and taller.

When Ray entered a room, the veterans lit up and the rookies sat up. Everyone wanted to be like Ray, and everyone wanted the respect he demanded when he strode in the door. He had been on the engine company out of Station 4 for 13 years.

One cloudy summer day, Jeremy, a young probationary member, came into the kitchen with bags of groceries and went to set them on the counter. Before he could, one of the plastic bags ripped open, and cans of tomato sauce and a few stray onions hit the ground.

Jeremy was a solid firefighter who tested well on the written examination and surprised some of the older members with his time on the physical agility test. He was the kind whose hair was always messy and whose shirt was never completely tucked in. Moreover, he had a kind face and was friendly. However, he was also an easy target because of his easygoing nature and his overall timid disposition.

"Shoot!" he said under his breath as he banged his knee against one of the cupboards trying to break an onion's fall.

As Jeremy was kneeling down to pick up the mess, another set of hands bent over to help. Ray knelt down by the kid and started handing him cans to put on the counter.

"Thanks so much," the 22-year-old said to Ray in a respectful tone, but avoiding eye contact.

"We haven't met," Ray said.

"No, sir," Jeremy uttered, taking the cans from him.

"Well, I'm Ray, and you are . . . ?" Ray asked, extending a hand.

"I'm Jeremy. Been on for a couple weeks."

"Good to meet you, Jeremy," Ray said, shaking the boy's hand and giving him a piercing look and a confident smile. "Would have said hi sooner, but I've been on vacation for the last few weeks. Wanted to get the boat in the water."

"Man, it's awesome meeting you!" Jeremy blurted out. He was starstruck and Ray knew it.

"How have things been going for you here? Are you settling in OK?"

Jeremy paused for a second, and Ray quickly picked up on his slight apprehension.

"Um, yeah, it's been pretty good. I think I'm learning, but it will take some time."

Ray could tell the kid was stressed. The lack of eye contact told him that something was on Jeremy's mind even though he wasn't saying what. Ray didn't want to pry.

"Hey, I gotta hit the bathroom, but when I get back I'd love to talk more," Ray said as he turned toward the hallway leading to the bathrooms.

Ray disappeared into the other room, and Jeremy mustered a nervous grin as he began to sort out the items he got for dinner that night with added pep in his step. As he put the Texas toast in the freezer, two midlevel firefighters stormed into the kitchen. Bryan and Cody had been in the academy together and formed an awkward friendship, along with an abrasive attitude around the station. Both were adequate but not exceptional at their jobs.

"What's up, probie?" Bryan fired off when he saw Jeremy fumbling with bags in the freezer.

"Hey Bryan," Jeremy said weakly.

"What are you cooking tonight?" Cody pried as he snatched a box of noodles out of Jeremy's hand.

"Uh, spaghetti and garlic toast," Jeremy responded.

"No meatballs! What's up with that, probie?" Bryan yelled, half joking.

"You have one job around here as our resident cook, and you messed it up!" Bryan let out a loud chuckle.

"Should I run and get some sausage real quick?" Jeremy asked, clearly flustered.

"Forget it," Cody said. "Just get the water boiling and shut up."

They both brushed by Jeremy and started rummaging through the pantry for bags of chips to tide them over until dinner was ready.

Around the corner, Ray had been listening to the interaction the entire time. When he came back in the room, he could see the obvious frustration on Jeremy's face and the disturbing pleasure the other two were taking in flustering him.

Ray decided it was time to act as he walked back into the kitchen. Bryan and Cody both greeted Ray as they grabbed their bags of chips.

"Hey, what's up, Ray?" Cody asked.

"What's going on?" added Bryan with a nod.

"Not too much," Ray said, eyeing a defeated Jeremy out of the corner of his eye.

"This kid forgot meatballs," Cody said with a nod toward Jeremy.

"I always used to forget something," Ray said.

Jeremy stood and stared blankly at the wall as he let the large pot fill with water at the sink. He had one hand on his hip as he steadied the pot with the other.

Bryan and Cody emptied snack bags of chocolate cookies into their mouths and threw them in the trash. They began to walk out of the room, but Ray caught them on their way out.

"Oh, hey, guys," Ray called to them.

They both turned and walked unevenly back into the room, picking their teeth and brushing the crumbs off their shirts.

"What's up?" Cody asked.

"I was wondering if you could do me a favor," Ray said with a poignant gaze at the floor.

"Sure," Bryan said, hiding annoyance.

"I was talking with Jeremy about going over some water-flow drills, and I didn't know if you guys could cook dinner tonight while we did. It would mean a lot."

Cody and Bryan looked at each other and then at Jeremy, who didn't dare make eye contact. The probie knew that the subject had never been discussed at all with the legend he met only 5 minutes prior.

"Sure, no problem," Bryan said with a tense face.

They both grabbed another bag of chips from the pantry and began to get things ready for the cooking they were suddenly assigned to do.

"C'mon, Jeremy. Helmet and gloves on and meet me out back," Ray said as he turned to walk to the bays.

Before he and Jeremy left, Ray stopped and turned toward the two firefighters in the kitchen.

"Oh, and Cody, would you run and get some sausage if you could? I know the other guys love some protein with their carbs, and I don't want them thinking we were plotting against them," he said with a devilish smile.

"Sure, Ray," a suddenly perkier Cody said as he grabbed the utility vehicle keys off the hook and made his way out the door.

Cody and Jeremy walked out to the bays. Ray, lagging behind them, stepped back into the kitchen for one last moment.

"Hey, Bryan?" he said lightly, glancing over his shoulder to make sure they were alone.

"What's up?" Bryan said as he began to chop onions.

"Really appreciate it. Didn't want Jeremy to be in here with us. He's trying to settle in, and I know he's nervous about being on the tip for the first time. I think some evolutions out back would do him some good," Ray said with a quick smile and a reassuring nod.

Bryan nodded with a sincere smile and a nod of his own. At least for that day, he got it.

Options

Many different roles are being played in the scenario above. Ray had the most options at his disposal because of his reputation in the department and the latitude his tenure and status afforded him.

By contrast, Jeremy had the least amount of wiggle room because his was a particular role most rookies were assigned. In other words, his role was by default, not by his own doing. Keep your mouth shut, do what you're told, and then eventually make your own way when you have a little time under your belt. We are instructed when we start our fire careers to be open to instruction, to be receptive to criticism, and to become information sponges to assimilate into the ranks as seamlessly as possible.

Bryan and Cody were caught in limbo. They had the ability to make changes yet chose to stifle that growth, for both themselves and Jeremy. The fire service has allowed traditions of department behavior involving the mistreatment of new cadets as a rite of passage for people who will one day be the face of the station. Verbal and even physical abuse is unfortunately common. The common justification provided for such rookie hazing incidents is that it is hardwired into the DNA of many departments. Such a rough-and-ready approach (larger departments may be especially rough) to the ranks toes a fine line between aligning with macho traditions and flat-out abusing a fellow firefighter because of personal insecurities and simply bad methodologies of instruction.

Throughout your career, you will arrive at crossroads that facilitate choices. Part of the balance is to make the best choice given the circumstances. Not just the best choice for yourself but for all parties involved at that particular time. From an occupational standpoint, we make these choices almost daily. We give the right medications for the appropriate ailment and the right nozzle for the appropriate suppression we need. But when it comes to making the right choices that will benefit each other, we tend to fulfill a role that we feel we need to play at that particular time. This can cloud our foresight of big-picture repercussions.

Forks in the Road

Let's focus on Ray again. Having accomplished so much in his career in the eyes of his peers, Ray could do just about anything he wanted within reason and probably get away with it. He could play the hero trump card whenever the mood struck, and most of the people he worked with would give him a pass.

In the scenario, Ray was at a proverbial fork in the road when he witnessed the verbal abuse the midlevel firefighters were giving to Jeremy. He essentially had three options, starting with the least desirable:

1. *Join in the verbal abuse alongside Bryan and Cody.* By adding his two cents into the mix, Ray would not only further drive Jeremy into the ground, but his comments in particular would be biting and taken much worse because of the credibility at his disposal. Although the midlevel firefighters' remarks are belittling and destructive, they are nothing compared to the potentially devastating lasting impression Ray could leave on a pliable young firefighter who already may idolize him.
2. *Directly challenge the two aggressive firefighters.* Ray could call them out, defend Jeremy, and remind them exactly how much crap indeed rolls downhill. Although this would champion the cause initially, the long-term effects of this would not solve the problem. On the contrary, it would exacerbate the situation when he's not around. Down the road, the firefighters he berated in Jeremy's defense could potentially take out their resentment and embarrassment even more on him. Would this truly solve the problem? Hardly. A teachable moment for Bryan and Cody would pass if Ray were to validate that type of behavior by doing it himself.

3. *Take the high road.* Not only did Ray spare Jeremy a moment of derision, but he also incorporated Bryan and Cody into the solution to the problem. By including himself in the group and putting himself on their level, he allowed Bryan and Cody to feel a sense of teamwork while still turning the tables on them and keeping them in check. He helped Jeremy by using time to teach him something that directly impacts his job and then made sure he let the other firefighters know that they were contributing to the younger firefighter's growth as well. A lesson is taught, nobody is broken down, and the potential for changed behavior out of Bryan and Cody is born.

To answer a question that may be on your mind, no, this isn't a life-or-death situation; and yes, technically this could be viewed as simple banter between firefighters in a firehouse. But little things add up. A simple interaction over dinner can suddenly become inflamed or can smolder and be remembered. The bricks in the wall are laid one at a time, and the more we practice finding the balance in every situation that we encounter, the better off we'll be when there are major choices to be made. Ray found the balance in the situation right away and implemented it admirably.

Ray is considered legendary because of the entire package—not just because of his heroism on the fireground. He successfully managed to avoid the "buts" in stories told about him. In other words, if he made choices on a daily basis that were out of balance, he would forever have an asterisk by the stories told about him long after he retires. A story about Ray saving two small children by pulling them out of a bedroom during a vent, enter, isolate, and search operation would lose its luster if everyone hearing the story knew him as a pompous jerk.

Roles

We all play several roles in our lives. The number of roles increases as we get older because we develop more relationships and take on more family responsibilities. The ability to be ourselves should not be clouded by the number of roles we play.

As a father, my role is different than my role as a lieutenant. Different obstacles cause different roles to manifest themselves. My role as a father helping my son learn how to drive is clearly different than my role as an officer ordering a transitional attack on a structure fire. But I'm still *me*. I have the tools I have developed over the years to find the balance in each scenario I'm given. Through maturity and experience, I have learned to cut through the fat of a situation and break down exactly what the main decision will be to achieve—or lose—the balance. If done correctly, we not only do the right thing but are actually compelled to do it.

A stark contrast exists between a true role we need to fill to achieve goals and one we play for ulterior motives. On the one hand, there may be a time when a natural leader is pressed into action by stepping up to abate a situation and protect his crew through strict directives. Command presence on scene is a terrific trait that serves us well in the line of work we do. The role presents itself, and we must do our best to find the balance in the decisions at that particular time. In retrospect, it was usually the right decision we made.

On the other hand, a role played for an ulterior motive seems contrived, weak, and oftentimes simply transparent. This is more apparent when we are young. As we limp through our scholastic career, a tribal mentality develops at an early age, and kids grow into different roles. When I was growing up in the 1980s, definitive lines were drawn between the jocks, nerds, burnouts, and other social cliques. This was cut and dry, and everyone strove to fulfill particular roles in the way they dressed and how they interacted with others.

As we age, it becomes much harder to punch through whatever facade people are putting on. The tough exterior often protects vulnerabilities that we don't want discovered. Many of us find that playing the role is the best route to take. Thus, there is a great deal of sarcasm, offensive language, mean-spirited remarks, and ultimately resentment toward one another because of the insecurities all of us have. It's never too late to stop and reverse the act we are trying to display, but it can be difficult.

One of the best ways to reveal when someone is compartmentalizing themselves into a role is to watch their interactions with others in different arenas of interpersonal communication. If you pay attention, you'll notice how certain people act a certain way with a boss, another with the public, another with their peers, and so on. By contrast, the ones who *don't* act are usually well respected and liked in the department. They are the same with everyone, and people tend to gravitate toward those individuals very easily. We all know someone like that.

We all have coworkers who have decided they are going to be consistent with everyone in a detrimental way. These are the ones who are never happy, always complaining, and never positive. Grudgingly, they get points for authenticity, but at what cost? They pigeonhole themselves into becoming malcontents who always have a problem with something and ostracize themselves from everyone else through their behavior. What a sad and lonely life. However, there is usually a cause that perpetuated this behavior. Therefore, even these tough cookies can be brought back to center—closer to the balance. Unfortunately, that process has to start with them, or energy and time will be wasted on someone who loves wallowing in negativity as an easy way out.

Roles are essential for us to have an identity and a feeling of self-worth. As we promote through the ranks and our career begins to take shape in our department, our roles become essential for operations to succeed. Someone who has been newly hired, on the one hand, has a set of standards they are expected to meet based on the role they were hired into. Senior firefighters, on the other hand, are expected to play a role of mentorship and are looked up to as examples of what is possible through dedication. Officers and administration both take on a new set of roles through the promotional process. Without roles in the fire service, we would all be part of a rudderless ship with no direction.

We will always hit benchmarks as we progress through our careers. Although the roles we are playing are revealed daily to people around us at home and at work, the *way* we play them is entirely up to us. Don't play a role at the expense of who you truly are. You're doing yourself a disservice when you do. Let the role come to you and unleash the qualities you have on that role to do it the best you can. Eat the role up and succeed at it no matter what it is. Don't fall victim to the role and how enticing the potential power play looks. The role will eat you alive, and coworkers and family will be able to feel the palpability of what you are trying to pull eventually anyways. Stay true to yourself and others will respect and like you *regardless* of the role, not *because* of it.

The following sections describe details to consider regarding the roles in your life.

Not Being Yourself Is Exhausting

Think about actors on the big screen. They put time into learning lines, moving certain ways, expressing foreign emotions, and so forth. Now imagine playing a different role for everyone in your life—strong and silent for this group, sarcastic in that one, the list goes on. Talk about depleting your energy and time; many people doing this are running on emotional fumes by lunch. When you balance yourself and stop fulfilling a role, you find life gets much easier. There is a lighter feeling that you enjoy no matter what situation you are in.

Classify the Roles You Play

The roles we play can be split into three categories. Two are determined externally, one internally:

- *Roles defined by responsibility (external).* This type of is what we frequently use in the fire service. It is a role that we are

expected to fulfill, to tackle tasks we encounter because of the role we have in relation to people close to us. Some of these roles may include our occupation or what our family dynamic dictates is acceptable. There are set expectations in these roles that must be fulfilled to be successful at them.
- *Roles defined by society (external).* These are the roles that we take on to adhere to proper societal expectations—from interacting in public places to waiting in lines, sitting in a classroom, and so forth. They are the roles that have been externally determined for us to cooperatively exist in social settings.
- *Roles defined by ourselves (internal).* These are the only roles we establish and execute to appease our own minds. We will go to great lengths to achieve these roles, because they are usually a fabrication of our own emotions and needs. These roles are by far the easiest to manipulate, and they are the ones we have the most personal leverage with. It is from this internal place where we usually begin to act as something we aren't. These roles are all about perception. Triggers for this type of role behavior include wealth, social status, fame, and our own insecurities with our true selves.

Consider How Your Role Affects Others

The roles we play have an impact on others. A good example would be a stern, workaholic father who is unapproachable and hard on his children. This has a direct effect on his children and how they behave around him. They may be hesitant to ask for help and scared that they will do something wrong. Even worse, they may think that the guiding force of their lives is doing things the correct way and they follow in his footsteps as they mature. Pay close attention and do your due diligence to discover what impact you have on those around you.

Positive Role Models Are Great

Actual role models are terrific. Someone to emulate and use as a mentor is worth their weight in gold. We tend to gravitate to the most accepting people around us, and unfortunately they are often the most negative people. This is because it is easier to criticize than to praise, and playing armchair quarterback is common in our profession.

In our story, which would *you* choose to be your mentor, between Ray and the other two members on Jeremy's shift? Me too.

Reflective Prompts

1. Think about the roles that you play in your life. Most of us can come up with at least a dozen roles if we think hard enough. How many do you play, and what are they? Hint: There are several we play on a daily basis that you may not even realize.
2. Which roles are most important in your life? Who is affected by the roles you play?
3. Is there consistency in how you fulfill these roles? Are there particular situations where you play a role a certain way because that is what the role dictates?
4. Are you attacking the roles in your life with integrity and self-awareness to recognize whether the role is winning the situation or you are?

4 The New Hire: The Future

We need to accept that we won't always make the right decisions, that we'll screw up royally sometimes—understanding that failure is not the opposite of success, it's part of success.

—Arianna Huffington

Ready or Not

Chris finished cleaning up the pizza boxes and took out the garbage from the kitchen as Charlie Company began to unwind in the dayroom after dinner. He had been on the suburban department for only a month. He was making a good first impression on his new shift by staying on top of his chores. Chris never knew when someone was watching and evaluating his enthusiasm for his new position.

The Fiesta Bowl was on, and the firefighters were getting into it. Chris felt as though he was catching a rhythm with the senior members, and his captain seemed to like him. The pizza was plentiful, and they had only five squad calls the whole day. His belly was full of doughy goodness, and the night was winding down. They had gorged like a den of lions and were digesting in recliners when the inevitable arrived.

Chris hadn't been exposed to any true structure fires yet. About 10 minutes after he had finished cleaning up, the distinct tone of a possible working fire came in, and everyone jumped into action. For senior members, the process was routine. The pump operator used a logistical check-down process to find the location and the best route to get there. The officer in charge had deciphered the initial reports. The ladder crew were primed to go. Even Chris's running mate in the back of the engine, Steve, had a systematic approach to radio checks, personal protective equipment donning, and which tools were to be grabbed on arrival.

Chris tried to stay the course, using the training he got from the academy. The basic skills he had been taught were painted with a rather broad brush to get him familiar with basic skills and operations. Still, he felt like he wasn't ready. He tried to put out the vibe of confidence without cockiness (which is the kiss of death as a new firefighter). He donned his bunkers and climbed in the back of the engine, and they were on their way.

The sirens screamed and Greg, the lieutenant up front, got another update from the radio traffic: It was a confirmed working fire. Chris's pulse jumped and he began to take a mental inventory of what he needed to do on arrival. *Grab the cross lay. Flake it out. Mask up. Hood after mask. Make sure you have water at the door. Wait for your interior officer. Find the seat of the fire. Be aware of smoke conditions you have zero experience with. Make sure you're on the appropriate nozzle pattern for the fire at hand. Be aware of victims. Don't open up too early, but don't wait too long. Turn on your radio—and for God's sake, make sure it's on the right tactical channel.*

"Tags!" Greg yelled from the front seat as he slipped his arms through his self-contained breathing apparatus straps. Chris and Steve handed a bunch of tags to the front.

Forty-five seconds out. Mutual aid units started chiming in with their progress toward the scene. Chris focused on the storage compartment straight ahead of him with a quick glance out the window at the red reflection of their lights on mailboxes lining the road as they whisked by.

Greg glanced back toward Chris, noticing signs of mild anxiety in the new recruit's mannerisms.

"Chris, when we pull up on scene, grab a cross lay and get it ready in the front yard. Steve, help Logan pull an LDH [large-diameter hose] for water supply while I do my 360," he said.

Chris took a couple of deep breaths. He had done this before in training. *Remember the basics. One thing at a time. Be deliberate but be cautious. Do your part and the team will succeed. Follow orders. Don't get in the way but don't sit on the sidelines.*

The engine pulled past the house which had clear fire and smoke showing from an Alpha-side bedroom on the second floor. Greg got out and gave his initial radio report.

"Engine 5 to dispatch. We are on scene of a two-story, medium-sized, single-family dwelling with fire and smoke showing from the Alpha side division two. Engine 5 will be establishing a water supply and will be pulling a cross lay for primary search and fire control in the offensive strategy. Engine 5 has command. Stand by for 360."

It might as well have been in Chinese. Chris jumped out of the engine and grabbed the cross lay. *Stay focused on your job. You have one job; don't screw this up.* He flaked the hose out in record time and pulled the tip to the front door. The radiant heat from the fire was already starting to make his arms warm as he reached the front door. He heard Greg completing his 360° evaluation over the tactical channel just as the ladder company pulled in, and the sirens from mutual aid companies were rapidly approaching. Man, that pizza had been a bad idea. Chris took a knee and began to check the buttons on his coat and then fiddled with the straps on his mask. Greg came back and knelt behind him, donning his own equipment. Chris hoped and prayed there wasn't an errant strap or a mask seal leak as he connected his regulator and Greg radioed for water. The line sprang to life, and Chris checked his pressure. They were good to go.

Upon entrance, Chris could hear his officer easing him forward and telling him to slow down. His heart was pounding as they made their way through the small foyer and up the large staircase leading to the second floor. Chris could hear the crackling of the contents in the room as they wound up the stairs and down the hall toward the bedroom.

"Second bedroom on the right, Chris!" Greg exclaimed as he patted Chris's shoulder.

The curved staircase seemed like 50 steps and the hallway about four football fields long. Visibility was still decent, but smoke was starting to pour out of the fire compartment. Chris got lower as the heat began to pick up intensity, and he could feel they were getting close. As Chris arrived at the door, he was greeted by a massive wall of fire coming from a closet in the bedroom to his immediate right. Laying partially on his side, he turned to get a better angle, pointed his nozzle, and pulled back on the bale. Nothing. Just a small sputter of water trickling out. *This wasn't supposed to happen!*

"No water!" Chris shouted. His tone was not one of disgust or irritation, but rather conveyed sheer anxiety and panic.

"What?" Greg yelled. "What do you mean, no water?"

"I got nothing!" Chris said.

"Back out into the hallway," Greg shouted back.

It seemed as though Chris waited forever for the next instructions. He was frozen, hanging on every word Greg would say. He knew there were white knuckles clutching the nozzle under his gloves. As if that wasn't bad enough, he could see smoke banking down as it poured out of the room into the hallway. Visibility was getting worse.

The room is about to flash over. But he waited. He had botched the one thing they had instructed him to do: put out the fire. All those hours at the academy, and this was the result: a loser who, despite having beaten 85 other candidates, couldn't even put out his first fire without a hitch.

"Back in you go! Hit it, Chris!" Greg yelled once again from who knows where.

Chris went back into the room and let it rip. His stream cut through the thermal layering of the room, and dark smoke bellowed out through the broken front window. He frantically sprayed in every direction. He hit pockets of fire at every turn.

"Hit, hold, and sweep!" Greg directed sternly but calmly.

Chris got to his feet once the temperature cooled and advanced the hoseline to finish off the closet. A day late and a dollar short. Another tap came on his shoulder. It wasn't Greg this time, but a captain from one of the mutual aid crews. *How had they arrived so fast? Where did Greg go? Why do I suddenly feel like I want to throw up six slices of pepperoni and sausage pizza?*

Chris was told to shut down his line by the mutual aid crew. *How humiliating. He didn't even know when enough was enough. Why had Greg, his lieutenant, abandoned him and left him alone? Where had Steve gone? Why did it take so long for a backup line to get there?* Chris was trying to put all these things together in his head into a tentative timeline when Greg popped his head in the door and motioned for Chris to come back out.

Outside, the crew went to rehabilitate by a cooler that had been placed in the front yard. Several other crews were waiting on deck and others were staging for their overhaul assignments.

Chris took off his pack and leaned on the bumper. He felt like he was going to be sick. He swore never to eat pizza again.

The truck company crew made their way over to Chris as he poured water over his sweaty head. *Here we go. What smart comments are they going to throw my way?* He braced himself for their inevitable criticism.

"Way to go!" The truck operator said as he patted Chris on the back.

"Nice save, rook!" Another chimed in.

How? Why? He screwed up and he was getting . . . this?

Chris took the compliments in stride with a tired smile and weak fist bumps. He turned and walked back over to Engine 5. Steve and Greg were standing and drinking bottled water while they bled down their regulators.

"Good job, Chris," Greg said. "You did exactly what you needed to."

"I felt like a complete idiot with no water," Chris answered.

"We fixed it. Technical difficulties," Greg said with a smile.

"Where did you go?" Chris asked.

"What do you mean?" Greg said.

"I mean after I hit the fire, you disappeared on me." Chris paused, finally able to reflect back on what had just taken place.

"I didn't go anywhere," Greg answered. "You just decided to stay a little bit longer. Except for some extension in the other bedroom, you had it knocked anyways. I let you stay and play for a few extra seconds until River's mutual aid engine came upstairs."

Greg paused a moment before continuing. "I was there all along," Greg reassured him. "You did just fine."

A couple shifts later, the department put together an after-action review so that improvements could be made and errors corrected. After the other shifts reviewed them, it was Charlie Company's turn to listen to radio traffic and watch footage. Unexpectedly, Chris was blown away: Not only did he put the fire out like he should have, but there was a good deal of information being relayed which he blew right by and missed in the process. The radio traffic was what shook him the most.

Chris had gotten so caught up in the moment that he didn't notice the interior command presence Greg exuded. Greg had been able to identify the water problem, back Chris out, call Steve to trace the line and find the kink causing the problems, order a second line by the mutual aid company to follow their line in, and transfer command to an arriving chief with a better vantage point—all in under 60 seconds. The rookie simply didn't notice anything happening because he was so wrapped up in what was going on at the tip.

From the after-action review, Chris drew three conclusions: One, Greg had his back as his officer. Two, Chris had to become more aware of the overall scene around him. And three, he had survived his first fire.

We've All Been There

Chris's first fire may not sound identical to many of ours, but we can all relate if we turn back the clock to when we were on probation. Remember when you are finding the balance as a brand-new firefighter that routines take time to establish. In the scenario, Chris was surrounded by routine. Each member had a unique routine as they headed to the bays after the tone sounded, from how they situated their boots and hoods on the bay floors to how they clipped equipment to their bunker coats. There are no definitive lines to the endgame of being ready. Still, we all need to be ready. For new members, it can be tricky because you have no routine yet. We get advice from older members and try to save time on our own. Nevertheless, the most crucial element of routine is experience, and we don't have any yet. It's OK. It's also normal.

When the fire was under way and Chris was taking his mental inventory of what he needed to do, he did an excellent job of quieting his anxiety and doing the task at hand. New members are often good at this. However, Chris also completely zoned out on everything else around him. New members are especially good at that too. He was so laser focused on the task that he was oblivious to the radio traffic and whatever else was happening around him. He didn't even notice the wheels in motion to settle the situation down in a timely fashion. Tunnel vision is something many firefighters experience throughout their careers. To break this habit, it takes practice and it takes time. Even on suburban emergency medical services calls, we are taught to stick to the basics. American Heart Association cardiopulmonary resuscitation guidelines preach to look past distracting injuries to deal with the life-threatening ones in the paramedic setting. There's a reason that is emphasized in many protocols across the country: It's tough to stay on track when life and death are on the line. Again, normal.

The passage of time can be dilated when you are in a structure fire and conditions are going south. On one of my first fires I remember donning my personal protective equipment and getting to the door, advancing the hoseline, knocking the fire down, and feeling exhausted by the time I was done. I felt like I had run a marathon, and it wasn't until I was recycling outside that I learned the entire process took less than 10 minutes. But we work with high intensity and short bursts of energy sometimes. We recycle for good reason: Firefighting is hard. Time passes incredibly slowly when you aren't used to the ins and outs of fire behavior.

As a new firefighter, you need to be cognizant that hasty decisions lead to poor outcomes. Find your balance in each situation and realize that each one has a problem to diagnose, a decision waiting to be utilized, and an outcome waiting to be discovered. Taking just a few extra seconds will serve you well. There have been new members in our department who have observed a critical factor on scene of both emergency medical services and fire calls that determined the course of action for the entire crew. It was because they found a balance in the situation and realized something success hinged on and brought it to the attention of those around them.

We have all been new. I tend to correct senior firefighters when they jump down the throats of junior ones. We were *all* new at one time. Realize this as a new hire. The testing and hiring process is there for a reason. It filtered you to a point where the city you work for is investing a ton of money in you along with a ton of trust. You have been selected by them to take care of the people who live in their city.

Keep lines of communication open with your peers and superiors. This is one of the most critical factors when finding the balance as a new recruit. Ask questions. Take advice gracefully. The delivery won't always be the most nurturing, but deep down everyone wants you to succeed and do well. Not only because you hold value as you move along in your career, but also because you're one more person they can trade with when they need a day off!

Then there is the social aspect. We are programmed as social creatures. New recruits quickly develop a tribal mentality with people they mesh well with. Two significant relationships tend to happen, and it is important to be aware of both:

- The first relationship that is usually forged as a new recruit is with a midlevel member who has been on a couple years. Usually this is because they have a connection by virtue of closeness in age. New recruits and midlevel members are within reasonable shouting distance of each other in interests, music, relationship status, and so on. The connection is one that creates more of a relaxed feeling, and it is a reason right out of the gate to look forward to coming to work. This is a human trait of belonging that is essential to finding the balance early on. Find someone and connect. It matters.
- A second relationship develops in the first few years on the job, and it is different but equally important. This is the relationship a junior firefighter discovers with a senior member. Many of these senior members enjoy the mentorship aspect of their journey through the ranks. They enjoy imparting wisdom and helping a younger accomplish the steps necessary to survive and eventually thrive as their career moves along. Find a solid senior firefighter to mentor you. They are different than officers in that they have the same set of tasks before them as you do. They can be excellent allies in everything around the station from cleaning to pantry shopping to fireground tricks of the trade. Senior members won't be around forever. But they do have lots of knowledge that will serve you well.

Now, a word of caution regarding both of these relationships. It is easy to decipher the dynamics of shifts, and it is extremely easy to form opinions and adopt ideas that are jaded and overtly personal. Be cautious with whose word you take as gospel. Everyone has an opinion, and misery loves company. You will encounter your share of negative people who love complaining about everything, and it is dangerously easy to jump on that bandwagon. Before you know it, you will find yourself complaining about practically everything that goes on and it will be primarily based on hearsay more than anything else. Be leery of those who talk about people rather than events. Ask yourself if the story being told is verifiable by facts or just a personal opinion someone has because of a snakebite suffered long ago. Be careful. It becomes problematic while you try to find the balance.

A Smooth Transition Is Key

Finding the balance at home can be a fairly smooth transition if you recognize that there is change coming for you. Most of the time, we have been working as a part-timer, a volunteer, or we've been gaining experience on a private ambulance company prior to a full-time offer. These are all excellent ways to get our feet wet and a taste of what's to come to a certain extent. We grind and grind, until the wonderful day comes when it all pays off with a full-time offer. Elation ensues—we did it!

Once you begin the journey of your career, you will most likely get a considerable bump in pay. Like many of us when we get those first checks rolling in, you may start throwing money around like you're a rock star who just scored a number-one album. We buy the cars we want, start looking at houses, and book trips. Before long, we are pinned to the job we just landed. Suddenly that paycheck is needed. A simple way to handle this is to have a plan in place prior to your hire date and stick to it. Budgets are key: Throw money into long-term savings as early as you can.

One of my mentors, Captain Rodney Meadows, put it perfectly one morning as we sipped coffee after shift. As he lifted his mug to his mouth, he muttered, "The two greatest powers on this earth are love and compounding interest."

Love that guy. His wisdom was direct and to the point. We can't replace time. This isn't a financial planning book, but start early on with your savings. You'll be glad you did.

Find your balance at home and make time for your family. Make sure your balance is found outside the firehouse. It's tempting to get lost in a passion when you start off. Often, at this time there is not yet a family being raised, and that's a huge asset when you're young and full of energy. Wait until you're old and out of energy—by which time you have about five times the obligations to contend with. Starting your life with balance in all areas is a huge advantage that puts you well ahead of the game as you get older.

Above all, remember that you will make mistakes along the way as you go. With both families, there will be bumps in the road to success. As in any career, the balance must be found and the process must make transitions. You will catch your rhythm and you will do fine. Just remember that the balance is always the goal in any decisions you make.

Final Thoughts

Keep the following important factors in mind as you start your career with a department:

- *Learn the culture.* Every department has a culture established. It is difficult to get a feeling for it by a ride-along, an interview, and a station tour. Ask around. Go with your gut. Don't be shy about getting opinions on your potential employer. It's a safe bet that they are doing the same on their end. Once you've accepted their offer and begin working, sniff around a bit and get your bearings. Each shift has its own flow and dynamic. You can't get balanced by swimming upstream. Take your time.
- *Prioritize your routines.* As I mentioned, you will become a creature of habit. The routines you establish need to work their way out in a pattern from the center:
    * The inner circle of your routines must start with the essentials to functioning well during emergent situations. How can you speed up your response time? What tricks can you use to help with patient care? These are critical.
    * The middle circle comprises tasks at the firehouse that are not emergent but hold weight with your peers. How do you clean up after meals? What mental checklist do you have during rig checks and supply inventories? Make sure you have a firm handle on the middle circle before you progress to the next.
    * The outer circle represents the routines that are just yours. This includes the way you decorate your sleeping quarters, how you make your bed, where you place your International Association of Fire Fighters (IAFF) sticker on your truck, and other minutiae that officers don't care much about in the grand scheme of things.
- *Protect the home front through planning.* As you start your career, have the foresight to prepare the people close to you about what is on the horizon. You are embarking on a personal journey that will make you and them very proud someday—but not every day. There will be peaks and valleys to your development as you pursue your balance. Keep your family in the loop and let them know what's going on with your growth. They'll appreciate the heads-up and will understand a little better when the seas are rough.

- *Honor your strength of character.* Strength of character is a huge plus for a new hire. Many times, it is misinterpreted as righteous indignation or someone who pushes back against unfair treatment. That is not what I'm talking about. What I *am* talking about is being powerful enough to realize that you will be challenged by difficult people from time to time. When this happens, lean on your strength of character to persevere through those tough times, and don't let small battles dictate an entire war. I'm not advocating for tolerating mistreatment. I'm cheering for the individuals who stand tall regardless of what is thrown their way. Tap into that energy. Realize that most crises go away by the next shift. If they don't, someone with a power play will sniff it out and help.
- *Don't let your enthusiasm fade.* When you start out in your exciting new career, the sky seems to be the limit. But as we start to find our groove, some of us decide that this is just the way things are and sleepwalk through the next few years. Stay proactive and strive to learn. The time to try different things and join different teams starts early. Do as much as you can while you are the pupil. Because you'll wake up one day as the teacher with several junior members looking to you for guidance. Make sure your cupboard of knowledge is full when they ask.

Reflective Prompts

1. One of the main components of finding your balance out of the gate is establishing a series of routines. If you are a brand-new hire or on the verge of getting hired, you definitely have areas where routine will apply. What routines bring balance to your life?
2. What are some areas where you think you will need to follow a routine? Think about your day at the firehouse—from the time you get up for work to the time you punch out. You'll be surprised at the number of routines you already have integrated into your day.
3. Have you made a budget or created any other financial framework for your life? How do you decide what to invest in? How do you treat yourself within reason?
4. Are there members on your shift who you have forged a friendship with? Are there candidates who you want to have as mentors? How would you start that conversation?

5 Midlevel Firefighters: The Present

Outstanding people have one thing in common: an absolute sense of mission.

—Zig Ziglar

A Perfect Storm

For the most part, the suburban fire department ran on routines and ho-hum emergency medical services (EMS) calls. However, every once in a while, a call would drop that would gunk up the works. The growing station had a minimum staffing of seven firefighters, and one officer was required to run the evening shift.

This particular evening was no exception. Billy was the lieutenant that night, and his crew were on the other side of town responding to a shortness-of-breath EMS call. The fort was being held down by four midlevel firefighters of whom Lexi was the senior member. She had been on the department for a modest 9 years and was followed by Trevor, Pete, and Garrett—all of whom had between 6 and 8 years of experience. Their town saw very little fire on a regular basis, and the department as a whole relied heavily on limited training when they could coordinate it.

At around 2100 hours on that crisp fall night, the tone went off for a possible structure fire on the west end of town in an old farmhouse. The four firefighters raced to the bays and began the process of getting out of the barn in a timely fashion.

"What are we taking?" Pete yelled across the bays to anyone within earshot.

Pulling on his bunker pants, Trevor yelled out as well, "Lex! Are we taking one or both?"

"Two on the ladder, two on the engine," Lexi shot back as she slipped her coat on over her hood.

Trevor jumped into the ladder truck and waited for Garrett to climb in. Lexi did the same on the engine, and Pete hit the lights as Engine 6 and Ladder 6 rolled out and headed west.

"Engine 6 and Ladder 6 en route to 9712 Colgate. Dispatch, please continue with first box mutual aid," Lexi chirped into the radio.

"Copy, first box being notified," the county dispatch shot back.

On the drive, there was time to consider the circumstances. Colgate Road ran along the western border of the town, and numerous farms unfortunately posed significant problems along the country passage. Hydrants were sparse because of the rural mixture of wooded areas and rolling fields. Water shuttles were a reality, and stiff winds came across regularly out of the west without much resistance. The whole area was as flat as a pancake. Until they arrived on scene, it was anyone's guess just what they would be walking into.

As the senior member responding, Lexi began to think—and overthink—the possibilities waiting for them on arrival. She knew that mutual aid was going to take a while to arrive. Moreover, with two of their adjoining departments being made up of volunteers, it was going to be dicey if this got rolling. Even though the adjacent volunteer departments were filled with terrific firefighters, mobilization would take time, and response could be hindered given lack of consistent manning at their station. She also knew that Billy's crew would be tied up for at least another 20 minutes, even if they headed right to the scene after their squad call cleared the emergency room. She had been on some fires but never in this capacity.

Deep breath, Lexi told herself. *See what you have and react. Almost there . . .*

Kenny and Loretta Miller bought the Clancy farm in 2004 and started renovations immediately. There were numerous challenges in the occupancy, and the century home was a ticking time bomb for a firefighter: very old appliances; know-and-tube wiring that had still not been completely replaced throughout the house; space heaters throughout to counter drafts from old windows; the addition the Millers put on the back of the house was lightweight construction added to the original clapboard; lots of junk in the attic; more junk in the basement; boxes were teetering everywhere inside the home, which was essentially a hoarding situation.

An engine, a ladder, and four firefighters without an officer made a final turn onto Colgate Road.

As Engine 6 approached, Lexi could see dark smoke coming from windows on the first floor out of the Alpha side. Delta seemed clean from a distance. They were in luck: A hydrant was about 150 ft down the road from the direction they were approaching.

"Stop up here and we'll lay in," she said to Garrett as she unbuckled her seat belt.

The house was set back about 100 ft from the road and had a barn and two detached garages. There were a couple older trees in the front yard and some farm machinery scattered around the backyard. A pickup truck and a small Toyota Camry were next to the house, and it looked like no lights were on inside the house.

Lexi jumped out and wrapped the hydrant before walking briskly toward the scene, making sure the large-diameter hose fell as it should. Garrett pulled the engine past the house and yelled back at her as he began setting up the pump.

"It's in the basement, coming out the Bravo side!"

Never been on a true basement fire. Stay calm and think.

The ladder arrived with Trevor and Pete. They pulled behind the engine along the road and began assessing as well.

"Is that in the basement?" Pete asked.

"Sure is," said Trevor.

Lexi walked a couple feet away from them and began her initial radio report: "Engine 6 and Ladder 6 on scene. We have a working fire in the basement, and the company will be pulling a line for quick hit."

Then Lexi turned back to her comrades on the scene. "Let's grab a cross lay and hit this thing from that Bravo window in the basement," she said.

She began to walk toward the engine when Trevor called after her: "Lex, you want to do a 360 while we get this line?"

Oh man, she thought. *He's right.* "Good call, Trev," she said as she moved briskly toward the house.

A steady wind was feeding a fire that was running away from them rapidly. A couple cars had pulled off the side of the road to start watching the festivities from a safe distance. She had an audience now.

The pump on the engine was ready, and the firefighters started pulling the cross lay. Lexi made it to the Bravo side, and it was really going in the basement. She proceeded to the back of the house, where the Charlie side looked like a junkyard. It would be virtually impossible to get a line anywhere back there without some creative flaking. She glanced up to the windows and could see no movement inside on the second floor. Smoke was beginning to fill the first floor, and she was having trouble making out the interior. The only thing she knew for sure was that it was a mess inside. As Lexi reached the Delta side, her heart sank as she looked to the second-floor windows, because she saw movement in one of the bedrooms.

"Command to dispatch!" she yelled.

"Dispatch. Go ahead, command."

"We have possible victims on the second floor!"

She ran back to her stunned crew members, who had heard the radio traffic about potential victims. "Delta second-floor bedroom window, we have movement inside!" Lexi shouted. "Let's ladder the window and try to get up there!"

"We have to hit the fire, Lex," Trevor said, pausing. "It's already pushed up the walls and into the attic."

Risk a lot to save a lot, she thought instantly. "We have victims in the bedroom that are still moving around up there. Conditions can't be that bad yet," she shot back.

"Where the heck is mutual aid?" asked Pete in a slightly panicked tone.

Lexi cued the mic again to get an estimated time of arrival (ETA) for the other responding companies: "Command to dispatch. Do we have an ETA on mutual aid?"

"Negative, two engines and a ladder responding with no ETA."

Suddenly another voice broke over the radio: "This is River Engine 3. We are 5 minutes out."

This will be a pile of nothing in 5 minutes.

Pete chimed in, "Let's go two and two. We have to get them, but we have to get water on the fire now!"

That plan made sense. Lexi nodded in agreement, and Pete went to get the ladder off the engine. Trevor and Garrett took the hoseline and commenced their attack on a fire that was beginning to rage in the basement.

Pete was back in 30 seconds with a ladder and an axe, and he butted the ladder after positioning it to the side of the window. As he did, Lexi noticed he put it on the leeward side, and she quickly corrected him: "Windward," she said.

Pete nodded and immediately moved the ladder. He gave Lexi the axe, and up she went to the top. Smoke was now starting to come out of the cockloft, and she could tell it was hot.

Keep calm. One thing at a time.

She knew time was running out for the Millers, and there was no way she could go more than a foot or two inside—if even that far. She swung the axe and broke the window, clearing it afterward. Smoke started progressing darker and darker toward her.

"Fire department!" she yelled urgently.

She almost lost her balance as a black and white cat jumped out of the window and into her arms, more clutching at anything than answering her call.

"Holy . . . It's OK!" she said as she came down and handed the cat to Pete. It bounded out of his arms and ran toward the barn. Lexi scurried back up the ladder, but dark smoke was pouring out of the window heavily. The heat was almost too much for her even through her gear. She fumbled for her radio.

"Command to line crew, how's it going there?" she asked.

"Hitting the basement from the Bravo. Little progress."

I've failed. I can't believe I've failed.

She paused and collected herself. The cause was lost. There was no time to hang her head. It was heavy fire, which meant it was time to start thinking defensively. As she descended down the ladder, radio traffic broke in.

Lexi and Pete rapidly walked around to the Alpha side, where they could tell the fire had vented itself through the roof. The entire second floor was rolling. They quickly confirmed on the Bravo side that all second-floor bedrooms were now involved.

"River Engine 3 to command, we are on scene."

Another voice broke in right after. Somehow the relief of it was more than she expected.

"Chief 440 on scene. I'll be taking command. Engine 6 command, meet me in the front yard for a face to face."

Lexi walked around the side of the house toward the Alpha side and was passed by River's engine crew, who were pulling a two-and-a-half to the Delta side and began breaking glass in the basement to start putting water on the fire. Meanwhile, as she approached her chief, Connor, another mutual aid crew was setting their aerial up after navigating it down the driveway.

The chief gave a smile to Lexi as she approached him, taking off her helmet in the process.

"Talk to me, Lex."

"Ladder 6 crew is around back trying to put water on the basement. We tried to make entry on the second floor on Delta, but it got too hot."

"Your crew accounted for?"

Lexi responded with her personnel accountability report (PAR): "Yes, PAR of two."

Connor cleared his throat and scanned the scene. He then took control.

"Command to Ladder 6, give me a CAN [conditions, actions, needs] report and a PAR level."

"Ladder 6 to command, still putting water on the fire, and we're having little effect, PAR of two."

"Copy Ladder 6. Redeploy your line on the Alpha side. Break. Command to River Engine 3, open your two-and-a-half on divisions one and two from Charlie."

"Copy, River 3 opening two-and-a-half on Charlie."

Connor continued, "Lake 2, let me know when your aerial is set up. We are going to open a master stream when I have confirmation."

"Lake 2 copies."

Connor turned back to Lexi while he had a moment. He could tell she was upset.

Here comes the tongue lashing.
"Great job, Lex," he said to her, with an extended fist for a bump.
"What? It was a complete mess."
"Not at all. You saved the cat."
"But the Millers . . ."
"Are in Niagara Falls," Connor completed. "I spoke to the neighbor who came out while you were saving Eddie. He was the only living thing inside."

Eventually, the situation leveled off. The other mutual aid crews arrived along with Billy and his crew. Operations went defensive and they were able to salvage some of the house. The Millers were contacted and, although distraught over the loss of their house, they were relieved to hear that at least Eddie the cat was safe and sound except for a couple burnt whiskers.

Lexi learned valuable lessons that day, and nobody was hurt. She did the best she could with what she had, and the crew with her did a great job of patching things together until help arrived.

After the fire, an after-action review was held. The officers on Lexi's department agreed that the middle-tier members on each shift would undergo training to gain confidence with taking control of fire operations.

Teamwork Matters

One of the greatest challenges that faces middle-tier firefighters is an identity crisis. As we progress through our career, we become vital cogs in the wheel of the department. Unfortunately, along with that comes added tasks and responsibilities that we aren't completely ready for because we simply don't do them much before.

In the fire that Lexi was forced to run, she was confronted with a boatload of problems that would cause trouble for even the most seasoned officers. The key point of the story is that she was surrounded by three other members with comparable experience and training, and together they all did their best. Although there were moments in their interaction when someone observing from a distance might not have been sure who was in charge, that's OK. What the onlooker would observe is four individuals working together to solve a problem. They had found the balance with help from each other.

From the beginning, the crew tried to communicate while in a tough spot. The four of them were caught at the station without enough personnel, and not only can that be stressful, but it can run counter to many simulations. Many times, the majority of tactical simulations assume ample staffing and a full complement of responding units. Rarely do scenarios employ a short staff and such horrible conditions. Lexi and her crew were thrust into a situation that forced them to maximize the options at their disposal on scene in a short period of time.

The balance in the scenario was multilayered. On the one hand, several small decisions had to be made, and the balance of each was apparent. On the other, there was an overall balance that had to be sorted out by Lexi throughout the call. Her struggle was to find the balance between technicalities and managing the crisis in the best way possible. Technically, she was in charge of the crew and the decisions being made within it because she had tenure over the other members. She had the final call, and it was her duty to make the best tactical decisions she could. Because she recognized that the other three members were there as resources and not a threat to her authority, she collaborated beautifully with them and took their input gracefully.

The crew did an amazing job under rough circumstances. There were many loose ends that could have been tidied up more efficiently, but they were able to function well enough to get things done on scene. Here is where the identity crisis usually comes in: Middle-tier members are caught between being those boots on the ground and being in charge. It is crucial to recognize that when these kinds of scenarios present themselves, the team mentality should kick in. That's mutual respect. That's the *balance*.

During the call, numerous details could have been improved. Lexi could have given a better initial radio report and been more assertive with command. She could have had a more definitive plan to attack the basement fire. She and Trevor could have been more on the same page as far as resource allocation, to balance out the rescue attempt with suppression operations in the basement. We can nitpick all day over such small details. But the reason the scenario was a success was because Lexi found the balance in each small decision and executed the best she could. Of course, her report wasn't going to be perfect. It's OK that she forgot to do her 360° evaluation. Routines take time and practice to become engrained. But when she was reminded of different aspects of the call by her counterparts, she seamlessly made decisions that kept the fireground tactics moving forward. That is a critical component to the balance.

What if the Millers had been home and were trapped in their bedroom? What if they had perished? Would this entire scenario have been a failure? Although tragic, the answer would still be no. Only blatant disregard for expected actions and subsequent applications would justify such harsh criticisms.

The crew acted as a team from the beginning and achieved their goal. Perfection in verbiage and error-free tactics would never warrant internal reviews for improvement. Very few structure fires have ever been completely error-free. They did great.

Around the Back Stretch

When we enter our middle years in the fire service—or any occupation for that matter—we take stock of several facets of our life. It is common for us to play the mathematics game, as it dawns on us that we are approaching the 12- or 13-year mark, or halfway through our career. Around this time we begin to get antsy about what direction we are headed in. Is this where we envisioned ourselves when we first got hired? Do we feel that we are progressing the way we wanted? How are we fitting into not only the department's plans but *our* plans as we creep into our 30s, with our 40s and 50s now on the radar?

First, realize that everyone goes through this mental inventory. Not only is it essential for growth, but it is incredibly healthy, keeping us motivated and in touch with our goals and passions for different pursuits. To keep the balance, we must refine and reevaluate our blueprint of where our life is in the service and in general.

Once we are middle-tier firefighters, we have established several features of our lives. Our routines are now in full swing. Our identity has been formed, and we are now associated with an opinion when other members or people in our lives mention us. Through hard work and authenticity, these opinions are hopefully good ones, and we have been behaved consistently enough that everyone in every situation in our life has the same opinion of us and what we stand for.

Around this time, we also start to zero in on special teams and personal agendas that will enhance our experience and overall value in the department. This contributes to the balance in that it starts to tip our job away from becoming stale. Many times, midlevel members will become critical of administration and decisions around the firehouse because they want change to be brought to them instead of initiating change themselves. Find a passion within the service that you want to explore. Once you start classes and focus on a certain area, it is crazy how you gain a renewed interest in your job.

On the home front, things have usually settled into a routine. Keep an eye on things, and make sure the routines aren't bad ones. If you have a young family, make sure you are taking time to be present with them. Many relationships have failed because of a lack of presence that demonstrated a shell of a human when they were around. Don't just be there, be present. One of the hardest things to do at home is ask for help or to lean on your significant other. If you were up all night, let them know. If you need to decompress from a gruesome scene or an emotional EMS run, let them know. The worst thing you can do is force your family to become detectives to figure out why their mom or dad has become impossible to deal with around the house. They will appreciate your candor, and you won't look weak when you do it—just much more intelligent and trusting.

One habit I used to fall back on was to become a martyr for the sake of my family. I would stifle emotions in a vain effort to score points on some level. My wife would ask if it was a good night or a bad one when I got home, and I would usually answer "It was fine" or "I got no sleep." I knew that if I said I had a quiet night, I was expected to be normal and carry on with my day; however, if I said I got no sleep, that was my ticket to get out of yard work or one of her family's parties (we've all done it, be honest). The problem with that approach is if it doesn't work, we're stuck. Now resentment creeps into the situation on one end or the other. Either you're upset because your partner is dragging you to a 1-year-old's birthday party or they are upset because you aren't going with.

How do we deal with these situations before they become a problem? By having discussions that address insinuations head on. Talk to your spouse. Tell them that there will be times when you get home when you will be exhausted and not much fun to be around. Let them know that you'll have moments when your eyes are sore and your body feels like you got hit by a truck. Tell them you will do your best to stick to plans you've made together, and ask them to be understanding when these situations arise. Discuss this stuff when the waters are calm. In the middle of a fight is *never* a good time to get a point across gracefully. Remember, they chose to marry a firefighter. And we chose to marry a human being who cares about us and who makes us feel special. Being a moody jerk because you were on fire watch all night hardly brings out the best in our loved ones. Realize that ahead of time and let them know the backstory behind your testiness.

The Forgotten Tool

A terrific tool that is underutilized at the middle tier is communication. In the scenario at the beginning of this chapter, communication was key for the members to have the slightest chance at success. As you have hopefully realized by now, communication at home is equally as important.

One of the pitfalls of communication in the firehouse is the *way* we communicate. A classic form of communication when we hit the middle of our career is through sarcasm. When strategically placed, sarcasm has a comedic effect that draws tons of laughter and makes us feel witty and sharp. But if we begin to use sarcasm all the time, we start to cut deeper and deeper with our remarks in an effort to be edgy and funny. Middle-tier members are usually the biggest offenders when it comes to this for one big reason: accountability.

When we are new members, we have not established ourselves and caught our footing well enough to have a sharp tongue around the station. We must earn our stripes before we fire off sarcastic remarks. Then, as senior members, we ease into a more mature age bracket, holding ourselves accountable and filtering what we say in an effort to maintain leadership in the ranks. Finally, as officers, it is ill advised to repeatedly hammer away at our personnel with rude gestures and biting insults to look better at their expense. Officers must keep a close eye on how they are interacting with the shifts they run.

Midlevel members are different. Outside of blatant disregard for safety or racist or other offensive remarks, these members basically have free rein to say whatever they want without repercussions. Think about it: midlevel members have experience, no loyalty to a gold badge, a huge advantage over new hires, and a platform to criticize anyone around them.

When implemented correctly, communication can't be replaced. It dissolves implications. It spells out a thought process that can be discussed and applied for the best outcome of the situation. Furthermore, it can help us to find the balance in our relationships. If you have a chance to reread this chapter, again note the underlying theme that effective communication is the key ingredient in the recipe for success in any situation we come across in the middle of our careers.

Final Thoughts

The following ideas encapsulate the essence of being a middle-tier firefighter:

- *You are the next in line.* As you forge ahead through the middle part of your career, keep in mind that you are the next in line to take over as a leader in the department. Slowly but surely, senior members peel off one by one. As you creep up the seniority list, you will have to decide which line you will be a part of. In one line, you will get better vacation picks and increasingly do less backbreaking work as younger members are hired on; you have an opportunity to do absolutely nothing to aid in their growth, and you can protect yourself and your future at the expense of the department, both literally and figuratively. By contrast, the second line offers you the chance to replace the mentors who came before you and create a lasting impression on the people who will look up to you for inspiration and leadership. With either line, you must look in the mirror and decide who you want to be.

- *Be proficient at all levels.* With years under your belt, you have a taste by now of all three levels of firefighting. Make sure you have a good grasp of all three:
  - By the middle part of your career, you should already be an expert on the level of new firefighters and their skills.
  - At the level you are currently on, you should be excellent at your job and know the skills appropriate for your years of service at your particular department.
  - Last, become familiar with and at least proficient at the *next* level above you on the ladder of shift members and officers. Get used to doing tasks you don't normally do. Try to practice running calls and being in charge of a scene. As we saw in the story at the beginning of the chapter, it matters when you are called on to do something you may not be familiar with.

- *Fine tune your home life.* Most middle-tier members have somewhat settled into a home life that may be transitional in the near future. Kids grow up and move away. Bills change. Our body becomes a little less responsive. Our 30s are usually upon us, and the 40s start to see even bigger changes. Don't wait for things to already be in place. Start when these changes are approaching and adjust accordingly so that you are prepared when they arrive.

- *Choose self-accountability.* There are many choices you will have to make, and earlier in the chapter, I discussed how middle-tier members sometimes run without the short leash of accountability the other levels may have. Don't take advantage of that. It's purely up to you how you choose to go about your business, but if you sell yourself short on self-evaluation, you fall out of balance in a big way. Suddenly things begin to falter, and trickle-down effects take hold. Pretty soon, you have relaxed into a decreased level of performance that has a bigger impact than you might have thought. You're better than that.

Reflective Prompts

1. As a middle-tier member, how effectively do you communicate? Do you feel you would be ready to step in if an officer was unavailable in a crisis? What kind of training do feel you still need to succeed as a stand-in leader?
2. Take a look at yourself and ask if you can do better. In any facet of your life, just work on making yourself slightly better at something. After doing so, ask if the sense of gratification is there now. Then move to the next facet.
3. What plans do you have for learning about the next level? Are you ready to step in as a senior member? Not only on a skill level but on an emotional one as well? If you feel out of balance with this evaluation, why? What is eating at you in certain situations that you know you can fix?

6 Senior Firefighters: The Past

I've learned that people will forget what you said, people will forget what you did, but people will never forget how you made them feel.

—Maya Angelou

A Peaceful, Easy Feeling

Chuck sat at the end of the kitchen counter and leaned back in his chair. He had transferred into Station 23 about 4 months prior and wasn't thrilled about it. After 22 years at Station 11, he had been sent to the west side to deal with a crew of middle-tier and lower-level members with a reputation of being loud, cocky, and generally a handful for the officers there. Chuck was 4 years in the Deferred Retirement Option Plan (DROP), and there were days he wondered why he kept going.

Along with about eight other people in the Western Hemisphere, Chuck still read the news in print. The Sentinel now came three days a week, on which days he would catch up on the news the old-fashioned way. With reading glasses on the bridge of his 59-year-old nose, he glanced up periodically from his paper to take a slurp of his coffee, which was black and steaming. His wife, Ann, had passed away 3 years ago after a valiant fight with cancer. She hung in there for about 13 months until her body succumbed. She made it 3 weeks past their 29th anniversary, and he considered that a parting gift.

Chuck had two kids. Lila, his daughter, lived in Atlanta with her husband of 4 years. She was a nurse at a children's hospital, and her husband ran an accounting firm. Chuck saw them about three times a year and knew it was tough for them to get away to see him. His son, Brendon, lived in Boston and did excellent work as a financial advisor. He would fly into town now and then so that they could catch up. They all had a solid relationship, but Ann had been the apple of his eye. When she passed, Chuck's plans changed.

Originally, Chuck wanted to go 5 years in the DROP to pay off the house and save for travel. However, when Ann fell ill, he decided to stay on because there wasn't really a goal anymore dictating when to retire. Still, the last thing he needed was to babysit young "kids," so he mostly kept to himself at the station. He was a fixture on Ladder 11 for the past 20 years and now was designated as the pump operator on Engine 23. He was pleasant enough to be around, reading his paper and listening to the Eagles when he could. The main thing Chuck realized at his new station was that he had become irrelevant. Sure, he played a role on structure fires and pumped proficiently enough. By and large, though, he had become obsolete in most other areas as the new blood took hold of the day-to-day activities of the station.

Chuck had worked with his lieutenant, Mark, at Station 11 prior to Mark's promotion several years ago. They were both on scene when the city lost three brothers in a floor collapse of an abandoned warehouse in 1998. They saw a lot of fire together and had gained mutual respect for each other in the process. Mark and the shift captain, J.C., gave Chuck plenty of latitude with training and duties around the station. Chuck would gracefully acknowledge their offers to ease his duties but always made sure to do his share.

Marginal respect was given to Chuck by the other shift members. Needless to say, their humor and interests didn't align with those of their senior member. Conversations seemed abrupt and forced. Everyone did their best to coexist with him, and he returned the favor. He tried to eat at the table for dinner with the shift, but as their crude humor and gutter mouths took a toll on him, he began eating alone in the kitchen when meal time rolled around. The officers assured everyone that this was a normal progression, and it really didn't matter to most of his colleagues—and especially to the "Frat Boys"—where Chuck ate.

The Frat Boys were a pack of four middle-tier members who went through the academy together and blazed their way into full time. The officers on the shift tolerated their antics because they were *really* good at fire suppression and got after tasks in an aggressive and proactive way. These firefighters checked every box in the public's eyes: strong, good looking, aggressive. They left some of the older members shaking their heads.

Chuck had adequate skills—as was true his whole career—yet knew he couldn't hold a candle to his coworkers physically. He shared with them the pride and intensity he once had for the job and felt it was his duty to nurse them along whenever he could. He had seen others start to wither on the vine as their careers progressed, and deep down it was refreshing to see the commitment and energy of the younger members.

A few incidents occurred around the station that a more abrasive personality than Chuck would have challenged. Rory, one of the more outspoken members of the Frat Boys, called Chuck out for being too slow to the engine on a car fire. Tommy and Rob both made cutting remarks when Chuck made an error packing hose after testing the large-diameter hose. Such remarks were usually shrouded in sarcasm—as is usually the case when someone wants to get across a point that they can't make directly.

Chuck took their criticism in stride and did his job. He was playing with house money at this point in his career, and he was generally too tired to verbally spar with the young members. Four months had gone by, and Chuck's two closest friends at the Station 23 were Don Henley and Glenn Frey.

Go Time

A fire broke out around midnight on a Saturday night, and Station 23 snapped to action, heading for the door. Four different onlookers had called to report heavy smoke and fire coming from the back side of a large house. There were also reports that the house was abandoned. Engine 23 and Ladder 23 zipped off the front pad, responding to the scene. The house was just seven blocks away, in a rundown part of the city. Chuck drove the engine around parked cars and was able to dodge moving ones. A junior lieutenant, Sean, from Station 18 was in the right seat with Tommy and Rory in the back.

"Step on it, old man!" Tommy yelled toward Chuck, half-kidding.

"I smell smoke. We're getting close!" barked Rory as he fastened the last snap on his jacket and spun around on his knees to see what they were driving into.

Engine 23 pulled in front of the house, with dark smoke coming from the Charlie side. Low hanging wires and parked cars were going to make aerial operations a problem, and Chuck pulled past the house, threw the truck into pump gear, and began taking a hydrant.

Sean radioed, "Engine 23 on scene. We have a two-story brick home with smoke showing from the Charlie side. Engine 23 will be in the offensive strategy and will be pulling a two-and-a half for fire control and primary search. Please continue with a first box and strike a second. Engine 23 has Carter Road Command. Stand by, I'll try to get a 360 report."

"Copy, upgrading your box," dispatch responded.

Scattered people gathered across the street as the crew flaked out hose and Sean briskly headed toward the Charlie side around Bravo. It was tight between houses. Although events were progressing rapidly, Chuck was able to handle the water supply fast enough. The ladder crew arrived on scene and started pulling a cross lay off of the engine themselves. Rob and Marcus (another of the Frat Boys) were lightning quick with their flaking and were ready for water in a flash. Sean broke back onto the radio as everyone donned their masks.

"Unable to complete 360 due to a fence in the back. But the fire is definitely back here. Get that hoseline back here."

J.C. was intercepted by a neighbor and took initiative to gather intelligence about any possible occupants. Meanwhile, Rob and Marcus were ready with the cross lay in the front yard and were getting fidgety.

"Let's go, Cap!" Rob yelled.

"We can pinch this thing out, Cap. Let's do this!" added Marcus.

They both knelt in the front yard and checked their hose. Marcus got up and checked the front door, which was locked.

Sean could be heard again over the radio.

"We've made entry on the Charlie side and have knocked down the fire. We are progressing with limited visibility for any extension."

Meanwhile, something wasn't sitting right with Chuck. He knew that the fire had progressed for a few minutes prior to their arrival, and given the construction of these houses, it was extremely likely that it had spread. He made sure everyone had water and turned back to face the Alpha side again. Although it was frowned upon, he walked away from the pump and took a quick look at Delta—call it a hunch. Right away, he could tell two things: First, there was smoke puffing in and out of the Delta side of the house, and there were dark smoke stains on the windows; second, this was clearly not one house but *two*.

Two more engines and a ladder arrived, and J.C. was called over by a battalion chief who had arrived right behind Engine 12. While J.C. was distracted, Rob and Marcus, left unattended, both began to approach the front door, this time with the irons. Marcus pushed the Halligan bar in the doorjamb and took a good bite with the adz of the Halligan. With one swing from Rob's flat-head ax, they were as good as in. Both looked back in J.C.'s direction and then at each other. They were weighing their options and decided it was time for the hero ball. Rob gathered his footing and was about to swing.

"Don't move!" a thunderous voice bellowed. It was Chuck.

It seemed like nobody had heard anything like it before. The sprawling anthill that was the fireground paused as if suspended in time. Everyone looked up suddenly, startled and puzzled. Chuck walked briskly over to Marcus and removed the Halligan tool from his hand. He removed it like a stern father who was grabbing a sharp knife from his toddler.

"Get back now," he said more calmly but still to the point.

Rob and Marcus retreated from the porch to the front yard with the mustering mutual aid crews. J.C. immediately made his way over to the front porch where Chuck, and the two stunned Frat Boys were standing.

"Captain, we've got stained windows and smoke puffing from seams on Delta. These are two separate houses, and this one up front is a pizza oven."

J.C. took all of 10 seconds to scurry to the Delta side to confirm the circumstances. He then hustled back to where the three waited.

"Jesus, Chuck. You're right. Command to interior," J.C. barked.

"Interior, go ahead."

"We have two residencies. Repeat, we have two residences. You are in the back house, do not breach any doors and hit hotspots from the Charlie entrance only."

"Interior copies, hitting hotspots only."

J.C. continued, "Command to Ladder 12."

"Ladder 12."

"Set up your aerial around these wires and let's get some vent holes in this front occupancy roof."

"Copy."

Just then, one of the windows on the Delta side blew out from pressure that had finally needed somewhere to go. Personnel scurried about and prepared to do a quick hit from the outside while Ladder 12 continued to prepare for ventilation.

J.C., Chuck, Marcus, and Rob backed away from the entrance as the house vented out the window.

"Good work, Chuck," J.C. said, turning around. But Chuck was already back at the engine, fussing with the water pressure from the pump on Engine 23.

J.C. turned to Rob and Marcus. They both were looking at the fire blowing out of the window. Rob had taken off his helmet and was running his fingers through his sweaty brown hair. Marcus slowly shook his head in disbelief. The magnitude of what just took place started to catch up with both of them. They could both feel J.C. and his glare.

"I don't know what happened, Cap," Rob finally said.

"Chuck saved both of your lives is what happened," J.C. quickly shot back. "You don't freelance and make decisions until we're ready, got it?"

"Yes, sir," they both muttered sheepishly.

Eventually, Ladder 12 was able to vent the front dwelling, and operations were able to move forward. In addition, Sean's crew knocked down enough fire in the back dwelling, and ventilation was successful there as well. The fire was knocked down by the interior crew, who made a quick hit on an Alpha window, allowing entry to be made after the roof was opened up. There were no injuries.

Back at the barn, once everyone showered it was well past 0400 hours, and the entire crew was exhausted. The trucks were put back in service, and some members tried to get a rest in the recliners while others toughed it out with a pot of coffee before the shift ended. The feeling was relieved and subdued compared to other days at Station 23.

Chuck sat by himself in a chair just inside the bay doors drinking coffee and listening to "Jessica" on his small speaker. He watched the light traffic creep by and the streetlights changing colors in rhythm to his internal soundtrack. Marcus and Rob came over and addressed him.

"Hey, Chuck," Marcus said as they approached.

Chuck leaned back in his chair and looked over his shoulder, eyebrows raised.

"Yeah, we wanted to thank you for tonight," Marcus said in a sincere voice.

"Dude, I'm so glad you told us to wait," added Rob.

They each extended a hand, and Chuck shook each of them respectfully.

"Gotta protect each other," Chuck said with a smile.

Marcus and Rob pulled chairs up with coffee of their own and watched the traffic with him, listening to the Allman Brothers Band for the first time in their lives.

The Relevant Past

Chuck's story provides a dramatic example of something that can happen to us any time, any place. The fire scenario was one that could have ended poorly, but Chuck drew from experience and applied it in a hurry when the situation called for it most. Leading up to that moment, he did his job and went about his business. He also refrained from disturbing the atmosphere the younger members had created at the station. That is a good senior member. This could not be a better example of someone who has found the balance in their new situation.

Senior members have stories to tell. Even though they may seem outdated at times, to varying degrees, such stories are relevant to what all members do daily. Remember, the fire service is steeped in tradition and lineage. There is nothing better than having a good story to tell. You can craft it however you want and embellish parts that need a dramatic hook (c'mon, you've done it, too). Nevertheless, the stories of senior members validate our efforts. We want experience and stories not only to model our own path to success but also to shape our identity in the service we belong to. We want to be accepted and part of something bigger than ourselves. The best way to learn how is to respect the people who came before us in the profession that we love.

Let this be a reminder on how senior members should be treated by other shift members. When they have strong character and good leadership qualities, they serve as great mentors. Occasionally, there may be senior members who have lost their way along the path to tenure, and such members can be more of an obstacle to growth than an asset. However, it is fairly easy to tell the difference.

Change

One of the hardest things a person can do in their life is accept change. Nobody likes change. In our career paths and our home life, change represents a real hurdle. Why? Because many of us have attached our identity with our situations at work or at home and have lost our true selves. Yet you're still *you*—always have been, always will be. You have just decided to make a pit stop along the way in your life to become a firefighter. But it isn't who you *are*. Rather, it's simply the situation you are in right now. Things can change rapidly, necessitating the balance and its powerful effect not only on our day-to-day lives at the firehouse, but our overall view of ourselves in retrospect. Once we recognize that we are ourselves—no matter what the circumstances—life becomes considerably easier, and the balance is found much quicker.

The reason that change is given such a prominent place in this chapter is its sheer relevance to senior members more than anyone else. Junior and middle-tier members are always progressing with change. They have more responsibility and fresh experiences awaiting them. While officers who have been on the job for a long time have transitioned to a new role in the department with a new set of responsibilities to keep them busy, senior members are in a unique situation where they run the risk of stagnation. They are not going to promote; they have seen plenty of situations; yet they are forced to change, and that can sometimes scare them.

We've all witnessed senior members who are ready to toss a new radio out the window or smash a new laptop computer on the ground. They aren't as savvy with technology as younger members. They can't see as well (I have readers myself). They monkey with new devices for cardiopulmonary resuscitation and gadgets for fire suppression. It's quite entertaining sometimes. But it is very frustrating for senior members to try and conform to things that are second nature to younger members of the department.

Culture Club

As senior members goes through the arc of their careers, they witness ebbs and flows of their station's culture. Not only do they witness the promotions and retirement of their mentors and coworkers, but they also watch junior members grow before their eyes.

One of the biggest challenges a senior member faces is similar to Chuck's situation in the scenario. Although rather introverted in his approach, Chuck was able to identify the balance in the shift, and he elected to take a backseat to maintain the cohesiveness of the younger group and their gung ho culture.

Many times, a culture is established by senior members and that culture is in turn passed down to younger ones. But as senior members get older, they begin to have less in common with newer members at the interpersonal level, and that can cause tension. The best way to combat this as you enter the back half of your career is to remember your interests, sense of humor, and insecurities when you were first starting out in the profession. Doing so pulls us back into the fold by realizing how we've matured and moved on to the next phase in our personal growth and development. If you don't think you've changed, just take a minute and ask yourself if you would let yourself as a 20-year-old be a parent to your kids or a grandparent to your grandkids. I didn't think so.

I didn't think I would be that person, but lo and behold, as I have aged, music has become louder, jokes have become dumber, and physical training has become more challenging. Remember, we all change as we get older—but the job doesn't. There are still myriad tasks to be performed urgently, and sometimes those emotional and physical demands can take a faster toll on us as we creep up in age.

Instead of entering a new era at the firehouse kicking and screaming, embrace the change. Recognize it. Realize that change is inevitable and that the loud music and big personalities are casualties of good firefighting and new approaches to old problems. They could actually make life easier on us as we see the finish line of our career approaching in the distance. This can be an enormous step toward achieving the balance as a senior firefighter.

Stay Fluid

An immense step toward success and happiness as a senior member is to realize the fluidity in everything as your service years tick into the teens and twenties.

At some point in our career, we hear senior members recount stories about how things used to be—most of the time with a story following immediately afterward. Younger members should take interest in these stories. This is not because they preserve a lost art of tactics that should supplant the latest technology. No, it's because you owe it to these senior members, and they owe you the stories they tell. Senior members are a wealth of knowledge and history that serves as a reminder of the legacies we leave and the lineage we've created.

Sometimes, a senior member will get bitter and become an adversary to his younger counterparts. Human nature dictates that as we get older, there should be a degree of respect that accompanies experience and age in an occupation. In our field, however, things are different. In no other field are we required to do virtually the same physical job at 25 as we are at 55 years old. Nevertheless, we are. Recognize this, and it will help you make better sense of things.

A senior member is going to have a senior moment from time to time. Accept this fact and it will make things less stressful. The old silverback of the clan eventually gives way to the younger gorillas at some point!

Keep the Play Pen Intact

Let's return to Chuck's scenario. As I mentioned before, he did a brilliant job of finding the balance within the shift he was part of, as well as the fire he was on. Things could have gone south rapidly at many different spots in the story.

In the day-to-day routine of the firehouse, Chuck could have asserted his dominance and started barking orders at the members of the Frat Boys, and he would have been backed with marginal support from Sean and J.C. because of his tenure. However, he realized that the best way to help these younger guys would be to allow them to grow and succeed at their own pace. Chuck recognized that these kids weren't deliberately trying to ostracize him as an outsider. It was just their way of breaking into the business.

The balance Chuck found between his home situation and work life was critical as well. With his kids away and his wife recently deceased, Chuck was battling through major transitions in his life away from work. Although such challenges sometimes point us toward dark paths of coping, Chuck kept alcohol and depression at bay with '70s music and current events. More important, he provided a steady environment for the young members. Believe it or not, the most abrasive younger members on the department are sometimes the ones who need the most nurturing. The senior member from Station 23 did what good senior members do: Allow the kids to play in the play pen and to spread their wings when the environment is least threatening.

When the fire eventually arrived in Chuck's story, he again excelled in his behavior and attitude. The younger members were gnashing their teeth in anticipation of the fire they were about to fight. Chuck once again stayed the course and allowed the kids to play. But, much like a responsible parent, when Chuck recognized the kids were going to stray too far out of the pen, he applied his experience and authority in a direct way to keep them out of harm's reach.

The consequences of inactivity could have been devastating at the fire. If Rob and Marcus had breached that door before Chuck identified the danger of a possible backdraft situation, we could be discussing how his visit to the burn unit in the intensive care unit went instead of reviewing a successful after-action review. Instead, good senior members use intuition to prevent catastrophic events from occurring. Sometimes junior members fall back on quick thinking to cope with rapid mistakes made. See the difference?

Which is the flashier news headline: "Two firefighters dive out of harm's way during blast from house" or "Two vent holes cut in a roof to let smoke and pressure escape before fire put out"? We all know exactly which story the public will tune in to and watch on the evening news. We also know exactly which one every incident commander prefers.

Another senior member with less experience might not have recognized the conditions in time. They might have been concentrating on the pump and had blinders on to the rest of the scene. There's no fault in that; the younger members would still have been at fault when they were freelancing by the front door. Nevertheless, Chuck kept scene awareness and was able to not only efficiently do his job but also fill gaps left by other members who had missed signs of danger. Well done, Chuck.

Final Thoughts

The following are key ideas that senior members should remember as they enter the golden years of their careers:

- *Accept change.* I've hammered this point throughout this chapter for a reason. It is critical that senior members keep change in mind. Procedures will be done differently, and younger members have a different brain that computes everything on a different level. They aren't stupid or spoiled; they are just different. Don't punish them with vindictive and frustrated behavior when society has crafted them in a certain way. If younger members are to give you the respect you deserve, give them the tolerance they need. Don't turn stale and bitter. Rationalizing that disapproval of new normals justifies being a jerk is false. You'll just be labeled as bitter.

- *Pass on tricks of the trade.* You don't need a formal classroom setting to impart wisdom. The best advice is often given during rig checks and tool drills. Two minutes of hands-on learning will make an impression on younger members that may last for entire careers. For example, if you see someone running into the same problems you once did when starting a saw, take a second to show them a method *you* were taught years earlier. You never know when you might be making a memory.
- *Gracefully transition.* Regardless of whether we realize it, we slowly transition in our careers from outworking problems to outthinking them. Our bodies slow down and take longer to recover as we get older. When we burst onto the scene, we are ready to chop, douse, cut, and breach everything in sight. The mystique dictates that we are a bunch of compassionate apes that tear through anything to get to our victims and pull them out of danger. Yes, our goal is to get to the people that need us. But as we get older, we tend to reach first for the handle before chopping down the door. Don't try to keep up with the youngsters. You won't be able to. Keeping your mind sharp is just as important as keeping in shape physically.
- *Find common threads.* Hidden underneath the brash veneer that younger members present may be common interests the two of you share. By showing an interest in their hobbies, you will probably establish a rapport with them that isn't as difficult to reach as you might have presumed. Music probably won't ever be on that list (it seems to get louder every year to me). Younger members have an appreciation for things you will never know unless you stop and get to know them. Remember, they are trying to find *their* balance as well, and defense mechanisms often start with insulating interests so that we don't feel vulnerable through disapproval.
- *Continue to grow.* One of the biggest things you can do for yourself is continue to learn. Just because you are in the drop doesn't mean you know everything. Keep soaking up the knowledge. It keeps your mind sharp, your career relevant, and the balance you have worked so hard to achieve in place.

Reflective Prompts

1. As senior firefighters, we are constantly trying to plot our career timeline in relation to our life timeline. Let's be honest, we're not getting any younger! What is your career timeline in the fire service? Do you plan on riding off into the sunset through the drop? Or do you still have work to do?
2. What do you want your legacy to be? Do you want to be known as someone who was relegated to the engine for the past 5 years? Or do you want to mentor junior members and leave your mark in a way that is fondly remembered?
3. If you were to leave today, what advice would give to the junior members who look up to you? Do they look up to you? If they don't, why not?

7 The Lieutenants: A Conduit Like No Other

The challenge of being a good manager is to get the best out of everybody, not just the few who are clones of yourself.

—Tim Field

Tough Decisions

Sitting back in his chair in the shift office, Cameron rubbed his bleary eyes. He had just finished combing through run reports from the day before to make sure the numbers matched on the various computer programs the station used. After a while, all the reports seemed to blend together.

Cameron had been promoted to lieutenant on Bravo Company just 6 months prior and was still finding his feet. The position had opened up because of promotions and another officer going out on disability unexpectedly, and Cameron was fortunate enough to score well on the promotional examination, edging out eight other candidates. The suburban station was moderately busy and consisted of 36 full-time firefighters including officers. He was the junior officer on shift below Jeff, the captain; and Matt, the senior lieutenant, who had been an officer for a little over 8 years.

There really weren't a ton of surprises thus far. For the most part, this was how Cameron imagined it would go: more paperwork, less lifting, a little more stress, and a different colored badge.

On this particular Tuesday, Cameron was stressed more than normal. Because of circumstances beyond anyone's control, he was the only officer on shift that day, and an incident occurred that wasn't sitting well with him. He had reprimanded one of the firefighters on the shift in front of the entire crew. This left an uneasiness throughout the shift that was festering. Normally, many decisions made on a daily basis might blow over, or maybe require a quick exchange after dinner to clear the air. However, this incident couldn't be overlooked because the altercation happened with the one person he would never have expected—his best friend on the department, Gabe.

Cameron and Gabe went through the academy together and had instant chemistry. Gabe was 2 years older than Cameron, and except for a brief 2-year period of shift shake-ups, they had been together on various shifts their entire careers. Their wives were friends. Their kids played together. They went on vacations with each other and attended parties at each other's houses. But today was the first day there was a clear disruption in the flow of their friendship.

The call in question happened early that morning. It was a motor vehicle accident (MVA) at a busy intersection during rush hour, and the typical chaos ensued afterward. After an airbag deployment, one of the occupants was banged up and trapped in her driver's seat. Although it wasn't the world's trickiest extrication, the problem arose because in Gabe's eyes, the police department weren't doing their job to control traffic, and incident management wasn't being done efficiently enough.

Gabe had grabbed the hydraulic spreaders off the engine given the damage to the driver's-side door, and his blood was starting to boil. One of the younger firefighters was struggling to find them on the engine, and in typical fashion, Gabe went to grab them himself. As Gabe rounded the corner of the engine, he was almost clipped by a rubbernecker who was not paying attention to the road. Although he wasn't hit, it was way too close for comfort.

Cameron was checking on the occupant of the second vehicle and only saw the tail end of the incident. As he approached the engine, he heard a heated exchange between Gabe and the driver of the vehicle.

"Watch where you're driving!" Gabe yelled into the driver's open window.

The driver went from distracted driver to offended citizen in a flash. The car stopped abruptly and the driver, a 40-something on his way to the office, slammed on his brakes.

"Excuse me? Who do you think you're talking to?" the driver snapped back.

Cameron quickened his pace as the altercation began to escalate.

"Pay attention! You almost hit me!" shouted Gabe.

The driver gave probably the worst reply he could have at that moment: "Hey, remember I pay your salary!"

"Worst boss on earth then!"

"Yeah, whatever. Guess the mayor is getting a call on this one."

"I'm just saying watch where you're going," Gabe barked again. He turned and walked to the vehicle, spreaders still in his hands.

Cameron arrived at the car and was able to get between Gabe and the driver.

"Control your people!" the driver barked before he succumbed to the beeping horns behind and the violent police motions ahead to keep moving.

"Sorry about that. Keep moving," Cameron was able to muster.

As soon as he turned around, Cameron witnessed a second situation developing. Gabe had handed the spreaders off to another member and was now engaged in an argument with one of the police officers near the extrication proceedings. Cameron hurried over once again with yelling in progress.

"I don't care what you were doing!" Gabe shouted in the officer's face.

The officer came up to Gabe's chin, and she was not having any of it. "You know what it's like trying to control traffic this time of day!" she shot back.

"Just get out of here. We'll just put cones out next time. At least they know how to slow people down," Gabe said rudely.

"You know what? I'm done," the officer said as she walked away.

A police sergeant on scene heard the commotion and headed over briskly from his police cruiser. Cameron cut him off before he reached Gabe.

"Sorry, Danny," Cameron said calmly. "I'll handle it."

"You'd better, Cam," Danny said as he straightened his collar and walked away to calm down his own people.

Cameron had enough. He turned back around and saw Gabe beginning to give one of the younger firefighters a hard time regarding his usage of the spreaders on the vehicle, being extremely critical.

"Gabe!" Cameron yelled.

Gabe whipped around, practically foaming at the mouth. He stared right at Cameron, wordlessly challenging his best friend of 11 years.

"That's enough," Cameron said, more calmly.

"I've had enough of this crap," Gabe said, beginning to walk back to the engine.

"Help with patient extrication. We'll discuss this later."

"Is that an order, sir?" Gabe replied sarcastically.

"If it needs to be."

Gabe walked back over, humiliated that four other members had heard the commotion.

The crew were able to extricate the woman from the car, and she was transported to the hospital. Gabe was in charge of the ambulance crew that transported her, and by all accounts acted in a professional yet subdued manner the rest of the way.

Nothing was said when all the units got back to the barn. There were a couple more minor matters to attend to around the station. Gabe and Cameron did their best to avoid each other, and the resentment was growing by the minute.

Here Cameron was, facing the first major test of his limited tenure as an officer. It sucked. He recounted the events from the MVA to critically assess if he could have done better.

Why was Gabe so upset? wondered Cameron. *He knew that behavior was not gonna fly. What the heck was his issue today? Should I have pulled him away from the scene to speak with him in private?*

Avoiding the fallout was not an option. Cameron knew he had to address this sooner than later but cleaned up his desk before calling Gabe into the shift office.

Gabe showed up a couple minutes after being called up front. He strolled into the shift office and plopped down on his chair, staring blankly into nowhere. Immediately, he dug into a defensive position, with his arms folded across his chest and his legs crossed straight out in front of him.

"What's going on?" Cameron started.

"With what?"

"C'mon, G. You know exactly what I'm talking about."

"You're condoning how incompetent our police department is on scene?" Gabe fired back.

"You know I'm not," Cameron replied.

"I almost got hit by a car, and all everyone is concerned about his how hard I kissed the driver's rear end."

Cameron sat back in his chair, and the two of them stared at each other for a few seconds before he reset his approach. "People are difficult sometimes," he finally said.

"Exactly. And I can't stand them," Gabe said. "It's like we have to bow to them, and then these cops show up and don't do anything."

"I get it," Cameron responded, "but we have to maintain our cool."

"Is this the part where you put on your officer's hat and start lecturing on good behavior?" Gabe snapped at him.

"I'm trying to talk this out. You know I don't want to go that route with you, but you're leaving me no choice here."

"Well, by all means, fire away!" Gabe shouted.

Cameron got up, took a drink of water and glanced out the window. He then sat back down and tried again.

"Look. I know your patience was being tried today, and I'm sorry. But what does it say to the younger guys when they see that?"

"I really couldn't care less," Gabe answered.

"We both know you do."

Cameron got out his notepad and began to write the date on top of it. Underneath the pad was an oral reprimand form that he was about to pull out. He didn't have any other choice, it seemed.

"Well, I have to—" Cameron started to say before he was cut off abruptly.

"Katie's got cancer," Gabe said in a much quieter tone.

Cameron stopped shuffling the papers and looked up at his friend. He could tell that Gabe was gut-punched by the ordeal and had to let out the news about his partner somehow.

"What?" Cameron asked quietly.

"Found out over the weekend," Gabe said.

"Jen doesn't know?" Cameron asked.

"Nobody does."

Cameron rubbed his face and slowly sat back in his chair. He could tell his friend was hurting.

"Been on edge a bit, I guess," Gabe uttered.

"Of course. I can see why now."

Gabe's eyes began to well up as his voice quivered slightly.

"We've been dating since high school, Cam. My head is spinning right now. It's so tough. The kids don't even know yet," Gabe said.

"Well, that explains a lot. We'll work through this," Cameron offered.

Gabe nodded. The anger had left his body and he looked defeated.

"In the meantime, I'll follow your lead as far as how you want to handle things here," Cameron said.

Cameron paused as the magnitude of the news began to sink in. After catching his breath, he continued: "Just let me know what I can tell the crew. Doesn't have to be specific, but they should know that something is up."

"I don't even know what to say," Gabe said.

"Whatever you're comfortable with," Cameron answered. "I'll put you on the pump for a while when you are here. The last thing you need is to deal with the general public right now. And with your blessing, I'll let Jeff and Matt know."

To this, Gabe gave an appreciative nod.

"If I get a call from the mayor about this morning, I'll handle it. You just take care of yourself and know that I'm here for you, like always."

The two firefighters rose and embraced. They had transcended the stereotypes of workplace conflict and bonded more than they had ever before.

"Do me a favor next time, though," Cameron said.

"What's that?"

"Come to me if you start getting a little wound up. We can take a break from this stuff when you need it. We're all here for you."

"Will do," said Gabe.

After Gabe left the office, Cameron sat back down at his desk. He took a long, deep breath before taking the written reprimand and filing it away.

A Delicate Dance

Cameron faced a situation no officer likes. In the scenario, he had to deal with multiple levels of stress arising from a single situation. First, he was a new officer in charge of the shift on his own for the first time. There is enormous pressure when that happens because we start to play the what-if game in our head. Many times, these head games snowball: Before we know it, we are wrestling with possibilities that are extremely unlikely to happen. In other words, our nature is to pressure ourselves to be able to deal with hypothetical situations that will probably never come to fruition.

Second, Cameron had to deal with one of his crew members directly violating policy on scene. Because of Gabe's actions at the MVA, Cameron's hand was forced. It was his responsibility to deal with a shift member who was clearly out of line not only with a member of the general public but also with the police who were on scene to help. To make matters worse, Gabe was starting to berate people on his own shift instead of focusing on patient care and doing his job.

Third, Cameron had to delineate between a friendship spanning over a decade and a duty to act as an officer on the shift. Numerous books have been written about the transition from coworker to manager. It's one of the most challenging obstacles a lieutenant will face. The problem is exacerbated when that friend is an extremely close one.

Finally, Cameron had to deal with Gabe's horrible news moving forward. He had to recognize that Gabe would be facing a long battle of emotions and inner struggles in the coming months. Furthermore, Cameron would have to deal with his own emotions.

In the end, an outcome was reached that maintained the balance of the situation. Although Cameron didn't have to dig too much into the root cause of Gabe's problems, it needed to be flushed out. Once it was, they could both deal with it in a civil manner.

What if Gabe hadn't readily disclosed his personal situation with his lieutenant at that particular time? More digging might have been in order. Cameron was ready to write Gabe up for his actions, and he felt bailed out by his friend's revelation. He certainly would not have been wrong if he wrote Gabe's initial conduct warning, but he would have missed the balance in doing so. It would have sat with him poorly, and deep down he would have felt that something was still out of place with his friend.

There's room for improvement in every situation. Give credit to Gabe for putting aside his anger, pain, frustration, and embarrassment. He owned up and told his friend what was going on, which potentially saved weeks of frustration, along with likely many more blowouts.

Two-Way Traffic

When we get promoted, one set of problems is replaced by another. One of the biggest problems we face after the gold badge is pinned on our chest is that we now must deal with both the messages from the boots on the ground going one way, and the orders from the top, going the other. This can be daunting, considering that we can do very little to change matters on our own.

At the level of the rank and file, it often feels like we are herding cats—with personnel angry about tasks and situations that arise on a daily basis. In response, we try to curb anger, manage complaining, and generally deal with the headaches that are always arriving at our doorstep. With little power to change matters on our own (except at the task level), we are stuck making promises to mention concerns at officer meetings to push them up the chain of command. Even though this may be frustrating, it's what we have been promoted to do. At these moments, we must find our balance and galvanize the troops into action. Keep your people in the loop and on equal footing. Make sense of decisions to ensure they will make sense to your shift members. They'll usually have a similar opinion.

From the top, we all know what rolls downhill. Lieutenants are the last stop on the assembly line of information processing before it lands in the laps of our members. Consequently, yours will be the face most often associated with bad news. Again, this is what lieutenants have signed up for. You must speak candidly and transparently, interjecting your own take on the new policy or rule without bad-mouthing the powers that be. Think it through before you zing the boots on the ground with news. There's usually a better way to spin what may at first appear to be bad news.

So what is a lieutenant to do? Recognize your role and accept it. Realize that you are indeed a funnel of information, and you *must* be someone your members can trust. Remember, you were one of them not so long ago. Put yourself back in the shoes you wore as a firefighter. Remember the concerns you had when you were given orders you didn't agree with and directives you felt were unfair. Be true to your abilities and trust your personnel the same way you want them to trust you. This will go a long way toward finding the balance with your shift members.

It's Just Different

When you are promoted to lieutenant, you have achieved two milestones at once. First, you have validated the long hours of studying and broken through to a new level of success. Be proud of that. At the same time, you have put a temporary hold on the plans your brothers or sisters who tested against you had as well. In other words, because you won the position, there are others who didn't finish first in the race.

As soon as you get promoted, make sure you address the membership either in one-on-one meetings or as an entire shift and explain to them how you are still a part of the same team they are. Nothing will upset people more quickly than a new officer who used to be their peer acting as if they are now the boss. Don't do that. Efforts to come off as confident and ready to lead can seem condescending and pompous; instead, be gracious on ascension to your new position. Start out on the right foot by communicating openly from day one.

No matter how much you may try, things will be different. Opinions will be formed, and you will be regarded in many ways while developing your management style. No matter how hard you show your commitment to still being part of the group, you no longer are. Conclusions will be drawn, and there is nothing you can do to get rid of *all* of them.

As an officer, you will have different tasks you have to complete. For example, you may have to put together training slips in the morning while the shift is washing rigs. If you don't convey that priority to them, an assumption may be made that you are now "too good" to wash trucks with them anymore. Not true: You just now have other commitments. Again, communication is the key. By no means am I saying to announce every single task you have throughout the day to validate your time to the troops, but keeping them in the loop as much as is practical truly does help.

Don't take offense that you will be treated differently. This might conjure thoughts of not being asked to play basketball out back or being left out of card games at night. That isn't necessarily the case. It might be quite the opposite: Because of your new position, firefighters may take on tasks for you such as washing the dishes or wiping down equipment. This may be uncomfortable for you, but it's their way of showing you respect. Take it in stride just as you would criticism. Thank them and move on.

Before you know it, you will have found the key to everything—the balance. It will take a moment to get your bearings, but it's there.

The Jack of the Trade

Consider yourself the jack in the deck of cards that is the firehouse, where the ace is the chief, the kings are the assistant or battalion chiefs, and the queens are the captains. In other words, there are a lot of number cards below you and several more important face cards above. You're going to have to integrate into a management hierarchy that is foreign to you. Procedures are in place for a reason, and regardless of whether you agree with them, you have to get a lay of the land before you can make individual strides to create change.

Just as new hires must find a rhythm and a routine in their daily lives at the station, so too must new lieutenants find their rhythm in the realm of management. Pick the brain of the senior officers. Talk to your chiefs and captains. Explore the newfound arena to which you have been introduced. Only through this method will you get a genuine grasp on the environment you are now in.

Don't let the frustrations of growing pains discourage you from your goals. There will be times when you don't give the best scene size-up at a structure fire. There will be others when you report to the emergency room with a critical patient who leaves out a piece of vital information. Stand tall as a human being, for crying out loud! It happens. Learn and keep moving. You may feel as though you are under a microscope, on display in front of senior members and management. In a way you are—but not in the way you think. If your senior officers are worth their weight, then they will evaluate you during your probationary period not on your mistakes but rather on how you adjust and learn from those mistakes.

When promoted to lieutenant, you adopt not only a shift but also a culture. The overall culture is established by the chief (for more on that, see chapter 9), but a more intimate culture is established by the shift captain (see chapter 8). After you find the balance in your shift and your duties within it, you can make your own assessment of what is already in place and introduce your ideas to the people who can get things changed. Until then, keep your senses sharp and feel the position you are now in.

If you fail to grasp the flow of the shift and start throwing suggestions haphazardly toward problems you don't really understand, your premature ideals will look like exactly that: premature. Nothing damages credibility like a young, uninformed officer who speaks up at his first officer meeting and reveals his grand plan for sweeping reforms of a department. The upper-level officers may be entertained, but they will hardly be impressed.

Wait in line until your turn arrives. You'll know when it has. One day at an officer's meeting, you'll spearhead a project or propose an idea that gains traction in the eyes of senior officers. This is truly a great feeling. While you wait, forge and strengthen your bond with your shift members. Nobody can make your life easier than committed, solid troops who want to be led by a confident and dependable officer.

Decisions, Decisions

Let's talk about one of the key balance points of this entire chapter—and essentially of this entire book. When you become an officer, you must find the balance in your decision-making while performing duties you have sworn to uphold to the best of your ability. While that may seem like an arbitrary statement that goes without saying, how you arrive at your decisions will make or break your success as an officer.

The balance that has to be found is the one that lies between making the right decision based on your own experience and reasoning and making the right decision based on the overall consensus of the players present. There is a duty to lean on our shift members to arrive at the best possible solution to a problem. Nobody is good at everything. Importantly, treating your shift people as pawns you strategically move around is far different from engaging them and allowing them to use their talents as part of a team. This can't be stressed enough.

As a firefighter, a clear delineation exists between being ordered to do something because someone said so and being given a directive that makes you feel like an active participant in the team's efforts to negate a situation. It is much easier to see this on the emergency medical services side of the ledger. A decent number of departments have paramedic crews wherein several similarly trained team members will go on the call. It would be foolish to bypass the opinions and input of other well-trained members regarding medical interventions that can directly affect a patient's outcome.

By contrast, fireground operations tend to be more chaotic. No two fires will be exactly alike. The situations firefighters face often rely on sound decisions made quickly by an officer who knows how to approach a specific scene (e.g., a structure fire) with clear directives. Even then, junior members may catch something the officer has missed. It is a clear sign of weakness to disregard opinions of subordinates. You don't have to agree with their assessments. It can turn into a teachable moment for them if they are wrong.

The balance entails treading that subtle line between asking your people for their opinion on a problem and being the final say in the problem-solving process. If you tilt one way too heavily, you become an "order barker" who tells his people when and what to do. That may work for a time, but concern will grow about your inability to listen given your hardheaded approach to solutions. Meanwhile, if you are constantly deferring to your people to make decisions for you, the tilt is the other way, to indecisiveness and a leader who lacks the self-confidence to overcome obstacles on their own.

When you meet in the middle of those embattled positions, you achieve long-term success. Make strategic choices based on knowledge and intuition—

but only after they have integrated your people and their strengths into those choices. Use the eyes, ears, and brains of those who are under you. They will thoroughly enjoy knowing their lieutenant is hearing what they say, interpreting their suggestions, and truly wanting them to be vital players in the overall plan of action.

Look back on your own career. Are you being the officer *you* wished you had when you were young and just starting out? Always try to be that officer.

Final Thoughts

Let's take another look at key points to remember as a lieutenant:

- *Communicate in an efficient manner.* Communication is an implicit key to success within the department. However, when I mention communicating in an efficient manner, this means finding the appropriate nuances that will resonate with the members *on your shift.* Each shift takes on a vibe of its own, and lieutenants must identify the best route in which to convey their messages effectively. A direct approach of setting expectations and following through is the default setting; however, staying aware of the way your people respond to directives and maximizing opportunities for individuals to respond effectively is critical for success.
- *Assure your people that you are an advocate.* The transitional period from firefighter to lieutenant is an exciting and stressful one. Still, as we settle into our new position, members will settle into their perceptions of how we operate. Stay aware of their concerns. Periodic one-on-one meetings are a fantastic way to keep up with what they have to say. Even if they don't have anything on their agenda per se, they will still appreciate the effort. Make sure that they are aware of your intentions to make their voices heard, their opinions matter, and the job they are doing is appreciated.
- *Trust.* When your job is on the line and you must answer to higher-ups, people tend to shrink their scope of trust. Sometimes it isn't even intentional; they want to know everything they can about each instance, so they are not left in the dark when questioned about tasks that were underperformed. Part of finding the balance within your shift is to establish trust with your

members. At the least, you will know their capabilities and what you can expect from them moving forward. Remember, members will do their best work with minimal guidance and maximum freedom when given a task. It is a vote of confidence you are giving them when they are assigned to do something that doesn't include you looking over their shoulder.

- *Review and improve.* Lieutenants must always look inward to ensure they are doing the best they can with the limited power they possess. They are part of the captain's shift, but that doesn't mean they should blindly follow along as an enforcer of sorts when they know something can be improved. Take mental notes on efficiency, and although you may have grand plans on improving the department, realize that you need to start somewhere: on the shift is the best place to begin. Know your captain and fellow lieutenants and strive for improvement.

Reflective Prompts

1. Take a moment to reflect on yourself and your career. What are some of the best moments you've had in relation to the officers you've worked with? What are some of more difficult moments?
2. Consider each officer you've had in your career. What was each one's best quality? What was their worst?
3. Take all the best qualities of the people you've served under and try to integrate those into your own management style. Are you already doing it? Were those qualities universal or unique to a particular officer?

8 The Captains: It's Your Shift—Sort Of

The employer usually gets the employees he deserves.

—J. Paul Getty

Out of His Element

People handle situations and stress in different ways. Most often, firefighters deal with stress by using a direct attack and a punch in the mouth. We know what the task is at hand, we're usually alpha personalities by nature, and we aggressively intervene to solve the problems facing us.

Sam had been a captain on his shift for only a couple months. He was very good at firefighting and had been a lieutenant for only a couple years before continuing his rapid ascension through the ranks. He had performed an impressive grab in 2016 and was recognized in an awkward ceremony with the mayor presenting an award honoring his bravery (although Sam never looked at it as heroic, saying instead that he just did his job).

Seventeen years on the department had flown by, and Sam was one of the best heavy-action firefighters the city had ever seen. He was a member of a midsized suburban department, and in response to coaxing from his officers at the time, he took the promotional examination and scored very well on those tests. Still, the captain's role was challenging for him. He was well respected by the members but felt that something was off. He had tried to put his finger on the cause but couldn't pinpoint the source until one hot summer day.

The truck company rolled to a stop in front of a two-story colonial on Deerfield Lane that had substantial fire on the first floor and smoke coming from the windows on the Alpha side of the second. The engine crew had arrived on scene and had grabbed a hydrant and pulled a cross lay for an attack. Sam was in a command vehicle that day while two lieutenants were manning the other first-responding rigs. Kurt was in charge of the engine crew and Pat was handling the truck company that day. Both were solid.

Sam heard Kurt's initial radio report, and the picture had been painted adequately for incoming units. He noted that mutual aid companies were steaming their way to the scene and Kurt was about to give his 360° evaluation momentarily.

"Deerfield command to dispatch," Kurt finally called on the radio.

"Dispatch. Go command."

"360 complete, this appears to be a division one room and contents fire. We will continue with offensive operations and make entry on the Alpha side."

Dispatch answered, "Copy: 360 complete, making entry on the Alpha side.

Sam broke in, "Command vehicle 27 on scene."

"Copy: 27 on scene," dispatch responded.

"27 to command. Meet me on Alpha side for a face to face prior to entry."

"Copy," Kurt responded.

Kurt briskly walked toward Sam and met him in the front yard.

"Hey, Sam. It seems to be in the kitchen."

"Go get it. Smoke coming from that second floor," Sam replied.

"27 to dispatch, 27 has command from Alpha."

"Dispatch copies: 27 has command Alpha."

Sam was antsy. Several aspects of the scene were making him squirm: The hose wasn't flaked out correctly in the front yard. The truck operator didn't place the rig at a good angle, so Sam knew that, unless repositioned, it would have to be short-jacked in a pinch. Nobody on the engine crew had grabbed a thermal imaging camera, even after he had written a grant to supply every unit with one. There was a possible exposure on the Delta side, yet there was no line pulled for protection yet. He gathered his equipment and headed back toward the front yard with his whiteboard. Sam watched as the engine crew went on air and began offensive operations. He arranged for the truck company to start throwing ladders and to handle utilities. He didn't even realize it, but he was pacing.

Two mutual aid engine companies arrived, and Sam assigned one to go on deck after pulling a second handline and the other to pull an exposure line for Delta side before staging as the rapid-intervention crew (RIC) company. He watched as Kurt and his crew made entry and began to blast through the first division. A mutual aid company captain strolled up and stood by to help Sam in any way he could.

"Hey Sam," the captain said with an extended hand.

"Hey, what's going on, Blake?" Sam replied as he shook hands and continued to monitor the scene.

Blake said, "I caught a neighbor on the way over. She said this house is being renovated, and nobody has been here for over a month."

"OK, good," Sam said. *That was a relief.*

Sam continued pacing, looking more like a nervous basketball coach on the sidelines than an incident commander on a fire scene.

"Why on earth don't they have someone shagging hose at the door?" he asked nobody in particular.

Pat came around to the Alpha side and joined Sam and Blake.

"Got the second floor laddered, and utilities are good," Pat said.

"OK, good," responded Sam, without taking his eyes off the house. "Why am I not seeing any progress here?" he asked Blake.

"They should have made it to the source by now, I should be seeing some better smoke someplace here."

"They've been in there for like 30 seconds," Blake said.

Pat chimed in, "I saw the neighbor. She said that they're fixing the place up and aren't here right now."

Confirmation of the first neighbor. Good.

"We aren't sure about that," Sam replied.

"Command to Destin crew. Grab a second line and prepare for search and rescue on division two through the Alpha side," he added.

"Destin copies."

Sam watched his orders followed and stopped himself short of grabbing the line himself. He then walked back to the front yard nervously and called to the other crew on deck.

"Command to interior. Can I have a report?" he announced.

"Interior to command, making progress into the main—"

There was nothing after that. Sam stopped and looked back at the house. *Wrong button hit? Trouble? What had just happened?*

Ten agonizing seconds later, with conditions worsening from the exterior, the Mayday was called.

Everyone stopped to get the traffic, hanging on every word.

"This is interior command, Engine 44's crew has a floor collapse, and we are in the basement! All three members down here, air at 75%!"

Sam started to run to the house but stopped. He grabbed his radio and composed himself before answering. "Copy 44. You have a collapse and are in the basement. We are sending a RIC for assistance. Remain in place and set off your PASS [personal alert safety system] alarm," he said calmly.

Sam then directed the RIC a few feet from him. "OK, they were straight back, but I don't need another crew following them down there."

He then motioned to Pat with a quick hand gesture. Pat and his crew were there in the blink of an eye.

"Pat, I need both the other handlines to knock the fire down. There might be some pockets with them in the basement, but we need visibility."

Sam paused and thought about what was going on. He then yelled after Pat: "And make sure everyone has their thermal imaging cameras on them when they go in. Move it!"

Pat and his crew quickly repositioned the exposure line, and one of the mutual aid companies grabbed the other cross lay. They opened both up and made good progress on the fire in a hurry. Sam began to walk toward the ladder but stopped short again.

"Command to 44, how are the conditions down there?"

"Not horrible, no fire."

Sam could tell the breathing on the other end was getting more winded. They were getting nervous.

"Pat!" Sam called.

Pat walked over with his crew still fully unloading a straight stream at the fire. Sam motioned to a mutual aid ladder company that had arrived. They walked up as Pat did.

"Pat, let these guys take over the attack. I need some eyes in the sky. Get up in your bucket and get over the house to start venting so I can see what we got. We're making some headway on this thing."

He paused again briefly.

"Command to dispatch, roll two more EMS [emergency medical services] squads here and upgrade our box."

"Dispatch copies command."

The one attack crew sounded the floor as they began to push into the house. The fire was being knocked back and Sam was happy with the results so far. He then radioed for Kurt again.

"Command to 44, how are conditions for you?"

"44 to command, we are OK for now, dark and wet down here but all members alert and accounted for. Air at 50%."

"Destin crew to command."

"Command, go ahead, Destin."

"We are in the hallway and have found the hole leading to the basement. We also found the basement stairwell. We will be protecting the stairs and the entranceway for RIC."

"Command copies. Break, command to RIC crew, follow Destin's line in and try to make visual contact with the downed crew."

"RIC copies, moving toward opening," the RIC leader said.

Meanwhile, Pat had his bucket lowering toward the roof line and could see what exactly was going on.

"Ladder 51 to command."

"Command, go ahead 51."

"We have a visual on the situation from above, there looks like some extension to division two but nothing imminent for a collapse."

"Copy. RIC did you copy that traffic?"

"RIC copies. We've made entry and have found the basement stairs. We will be entering the stairway for rescue operations.

"Command to 44, we have a RIC crew entering the basement. Stand by for contact."

"44 copies. Air at 15 hundred psi for all members here."

Another brutal 30 seconds passed before the RIC finally broke in again.

"RIC to command, we've made contact."

"Command copies."

"All members ambulatory and will be led out with the RIC crew."

"Command copies, exiting shortly," Sam said before continuing.

"Command to EMS units on scene, move up with equipment for evaluation upon exit of RIC crew."

"EMS copies, moving up."

Sam and his exterior crews waited until the RIC finally emerged with Kurt and his crew. They looked tired but they were alive and walking. One of the younger crew members was extremely exhausted and fell to one knee while exiting.

A collective sigh of relief reverberated through the fireground as it became apparent from the personnel accountability reports (PARs) coming in that they would all be OK.

"Command to interior crews, back out of the occupancy. We have PARs from all crews."

After the last crew had walked out to the front yard, Sam let it rip.

"Command to Ladder 51."

"Ladder 51, go command."

"All crews confirmed out of the building. Open up the master stream and hit any hotspots you see."

"Ladder 51 copies."

Pat and his crew demolished any pockets of fire that were left, drowning any possibility of the fire getting the upper hand again.

Still, Sam felt an anxiety like none he had experienced before. It was an overwhelming sense of helplessness and inactivity over a situation he would usually be able to handle even asleep. *Why was he so nervous suddenly? Was he getting soft?*

After the dust settled, Sam was able to see with the investigators what had happened inside. It turned out that part of the renovation to the house was due to some flooring that had to be removed due to some pipes that needed assessment. There was a gaping hole at the base of the stairs leading to the second floor, and the momentum Kurt and his crew carried caused them to go into the basement. There were obviously tactical learning moments regarding hose advancement throughout the event, but everyone got out safe.

During the next shift, Sam sat in the shift office and spoke with one of the assistant chiefs about the events on Deerfield Lane.

"Good job on that one, Sam," the assistant chief said.

"Ugh, that was tough," he replied.

"Tough sitting on the sidelines, isn't it?"

"Sucks."

"Well you did what you had to, and I would have done the same thing."

"I guess. You think I'm still cut out for this stuff?" Sam asked.

The assistant chief leaned on his desk and looked directly at Sam while replying: "Absolutely."

Sam leaned back in his chair and questioned the comment with his look.

The assistant chief got up slowly with his arms folded. He spoke after a deep breath: "Big lesson the other day, Sammy. You can't do everything. Not everyone will do things the same or as well as you did in their shoes."

"I found myself pretty wound up there," Sam replied.

"That will pass as you move forward."

"But do you think I was better suited to just be a firefighter?" Sam asked.

"Not at all," the assistant chief responded. "You just became a captain. You ran things perfectly and nobody was hurt in a bad situation."

"Yeah, but little things were driving me insane throughout the whole incident," Sam said.

"And they will. Welcome to management. You were able to tactically direct six different crews while keeping as cool as a cucumber," the assistant chief said.

Removing his baseball cap, Sam rubbed his head in silent consideration.

"The biggest question you have to ask yourself is, what was your effect on the fire? Would you have had as big an effect if you were, say, butt man on an exposure line?"

Sam slowly replied, "I guess not, if you put it that way." He got up and put his hat back on. "Thanks," he muttered on his way out.

The assistant chief went back around his desk and took a swig of coffee. "Leave the door open, will ya?"

"Yes sir."

"You'll be fine, Sam."

"You know, I think you might be right."

Sam turned out the open door and headed toward the bays.

A Different Kind of Transition

Sam's epiphany after the incident is a common realization when you are promoted to captain. The story had a good outcome, and the events that occurred kept the story interesting and relevant to what many of us fear when we hear the word "Mayday." But the events of the fire themselves weren't the main point.

Throughout the entire call, Sam was chomping at the bit to do things, yet he had to keep himself in check. Captains sometimes forget that they will be exterior command more often than they will be on an interior crew fighting fire. At several instances throughout the operations, Sam went to pick up a tool or flake out a hose but had to stop himself. This instinct wasn't a knock on his firefighters: He just had been a firefighter for so long, it was in his nature to act instead of direct. As our rank climbs, so does our scope of responsibility.

When Sam was a young firefighter cementing his legacy as a great team member, he relied on his skill set in the literal heat of battle to define him. Unfortunately, the same actions that would define a terrific firefighter would destroy the reputation of a captain, as he would be labeled scatterbrained or a micromanager. The lesson Sam learned is the same one many of our brothers and sisters do once they begin to run their own shifts and scenes—namely, we need to relinquish our role as a critical team player and adopt one as a critical team *leader*.

We live in a world of transitions. Regardless of whether Sam realized it, he was actively finding the balance as he ran the Mayday scenario. Think about if he had veered off course and began to help flake hose or took it upon himself to grab the RIC bag and set it at the door. Immediate results may have been achieved a little quicker, but the balance of the situation would not have felt even. He would be playing out of position, and the underlying feeling on scene would have reflected that.

Unified command on the fireground is essential for success. It is not there as a suggestion, but rather is a necessity for everything to work fluidly and with minimal interruptions in the flow of the work being performed. Sam was jittery at times, and his anxiety levels ebbed and flowed as the scenario played out. Overall, though, he kept the balance, and the entire incident was handled without serious injuries.

If you are a new captain, it is extremely important to embrace the changes of your role as the leader of your own shift now. Recognize that there are new tasks to conquer, new relationship parameters to be established, and new cultures to call your own.

A Culture of Your Own

As a shift captain, you have been saddled with a great responsibility. After you are sworn in and take the oath for your promotion, it is time to roll up your sleeves and realize that you have been given a newfound power to control the culture on your very own shift. Finding the balance here is extremely important. Why? Because moving forward, you are going to set the tone for how your shift operates and the overall mood of day-to-day operations, including the trajectory of morale and the trickle-down effect of your general approach to the workday.

When a captain is put in place, shift members become anxious. They are not sure what the culture will be with a new sheriff in town. In fact, most of them actually want guidance and acceptable parameters in which to work. There are very few success stories regarding shift development and advancement that involve one of the two polarities when it comes to management. Many of us can relate to one of the two—or have at least heard of them.

The first polarity is when the inmates run the asylum. Basically, this occurs when a new leader comes aboard and, to be one of the team, lets down their guard in an effort not to upset anyone on shift. After a while, the shift begins to take the cues from vocal senior members. Good or bad, after a while, the captain has undermined their own authority, and the shift spirals out of control. It becomes exceedingly difficult to get a shift back on the rails without major blowouts and a ton of additional work. The members may be familiar with you and your tendencies but never in the capacity of their captain. You must make a good first impression, and that takes planning. Think about your vision for the shift, and act proactively. Don't wait for the day of your first shift to formulate a pathway. Set rules in your head and stick to them.

The second polarity is the hard-line approach. Although not as difficult to recover from as the first polarity, this can also be challenging because the desired effect is rarely achieved. Instead, this approach is often met with resistance, and usually strict adherence to new policies for the shift is simply unrealistic; consequently, goals go unmet. This polarity is met with pushback, resentment, and ultimately challenges to authority—often in passive-aggressive ways. Before long, there is muttering in the corners, and members scatter when you enter the room. Except for a select few members who thrive on conflict, you will be avoided like the plague, and although your ego will be fed initially, it's lonely at the top.

The best way to find the balance when establishing your shift's culture is to think back when you were a firefighter and ask yourself what officer you wish you had then. Be that captain. Find the balance in everything you want to bring to your new shift. Ask yourself if it makes sense to you and to your lieutenants. Meet with them and see if they concur. Be open to change and to suggestions. At the end of the day, you are a team. But it starts with you.

One of the best ways to determine your plan is to get to know your shift members. In many departments (especially large ones), it can be extremely beneficial to sit down with each member individually. You can get a good idea of what they are about, and vice versa. At the same time, ask them about their concerns and personal goals. Not only will this preempt the initial guessing game about who you are now working with, but it can also dispel any preconceived notions the members might have about you by setting the record straight from the start of your working relationship. That's huge.

Moderation is key in almost everything. Your management style will have nuances that make you unique. As long as they feed into the endgame of finding the balance, the shift will flourish. There will always be bumps in the road to a successful career with personality conflicts—we're all human, after all.

Play the Game

You've done it. You studied hard, made lieutenant, then studied just as hard again, and now you're a captain. Your reward? Frustration at times. One of the major adaptations as you grab the reins of your shift is to take the long, wide brushstrokes used by your chiefs and condense them down into nice, neat instructions for your shift. Plenty of protocols will be discussed, and rules will have to be implemented by yourself and your lieutenants. Yes, it can definitely be frustrating, to say the least.

One great motto that will forever be relevant to a solid relationship between management and employees is: "If it makes sense, people will buy into it." This holds true from the captain's position all the way down to junior firefighters. In my opinion, this breaks down into something like this:

- Junior firefighters will do something they don't believe in because they have no other choice.
- Middle-tier firefighters will do something they don't believe in with some pushback among each other and some comments, with little influence otherwise.
- Senior firefighters will do something they don't believe in with stronger pushback and heavy influence.
- Lieutenants will do something they don't believe in because they have been instructed to do so by their captain and they want to stay unified among the officers but may push back with opinions.
- Captains *must* push back on something until they agree it makes sense. If the captains fail to push back and ask questions regarding the validity of the protocols, procedures, and rules that are to be implemented, the entire system breaks down. Pretty

soon there is a series of links in the chain that are subjected to things that don't add up. Eventually, the entire process begins to crumble because the captains failed to dig through to the core of the decisions being made. *Pushback* should not be confused with insubordination. In this context, questions will be asked, and these will be kicked off with *why* when personnel scratch their heads about a key decision made at the top.

Captains must realize that the chiefs and other decision-makers have agendas and priorities that may lose sight of what's best for the members of the shifts. This is not intentional; they simply have a different set of objectives on their minds. Therefore, it is the captain's duty to firmly keep their shift's best interests in mind when decisions are handed down. With a conscious effort to find the balance, even taking the smallest stance will pay massive dividends down the road in the forms of trust, camaraderie, respect, and long-term appreciation.

When you advocate for the troops through your actions with administration behind closed doors, you center your shift's balance without needing any directives or hard-line rules. Before you know it, admiration for you grows as you win over the members of your shift because you display empathy toward their daily struggles, even those you may not think are particularly relevant to firefighter performance. Behind closed doors, the word will get out to everyone that you championed a cause or stood your ground. The information passed along the pipelines at a firehouse can get out of hand and exaggerated very quickly.

Ultimately, captains must find their balance in unique areas from which other members of the chain of command are insulated. They also have a wonderful opportunity: A captain can craft a shift that has cohesiveness, professionalism, fun, and productivity that no other player in the game can match.

You Reap What You Sow

As a captain, your main goal is the same as that of virtually every other member on the department. We all strive to have a harmonious work environment where members feel appreciated and where their efforts matter. The rough and tumble days of hardcore paramilitary organizations are still in existence, but primarily limited to small pockets of large municipal departments. The job is evolving into an atmosphere of courteous professionalism for many houses across the country. The days of humiliating hazing and abuse as a rite of passage into the fraternity are fading away. The captain of the shift is the captain of the ship and must lead the way, as well as set an example through their own actions.

Being a captain is always an evolving process, given the fluidity of morale and commitment, which change almost daily. Don't become complacent with your technique. Once you have established the baseline expectations for the members of your officer staff and the firefighters under you, continue to monitor their behavior. It is easy to get lulled to sleep by a lack of communication. Something could fester for weeks or months on the minds of your people that could have been avoided with a simple talk.

It may seem daunting to stay on top of everything all the time. But staying on top of the members—including their insights, concerns, and goals—goes a long way. Before you know it, the shift will be running smoothly, the crew will be enthusiastic about their performance, and the long-term benefits will become apparent and be fruitful. Use the talents of your lieutenants and keep them in the fold. Loop them in regarding your goals and your vision for the future.

Don't be mysterious with your objectives. Although you may get passing admiration from someone here or there, it's counterproductive to cultural shift success. Above all, be transparent, friendly, assertive, good-natured, and steadfast with your demeanor. Ultimately, that combination will resonate throughout your entire shift in the long run.

Final Thoughts

Start with these suggestions to achieve the balance you need to succeed with your shift:

- *It had better make sense.* One of the keys discussed in this chapter is how directives pushed from the top down to shift members must make sense. Going over directives so you understand not just the what but also the why is huge. It is much easier to relay a new policy in the organization if you fully understand the reasoning behind it. Once *you* understand it, you can then formulate a way of conveying your point to the shift in the best way possible.
- *Your lieutenants are there for a reason.* A good working relationship with your lieutenants will go a long way. They are there to ease the burden of your job, not to undermine your authority. Make sure you all have the same vision for what you want to do with your shift and have them loop you in regarding

occurrences that might otherwise fall through the cracks. This starts with solid updates and communication. They are going to be a sounding board for the shift members, and everything should be discussed so that no surprises pop up out of nowhere.
- *Be steady but responsive.* You are performing quite the juggling act as you run your own shift. Even if you are a seasoned captain and have run your own shift for an extended period of time, don't assume that what has worked for years will continue to work. Tendencies change. Members retire. What may have been instituted a decade ago with other members may have to be tweaked to solidify the incoming core players on the shift. For example, you may have a new senior member who has taken the reigns after a retirement and has a completely different subset of strengths than the senior member preceding them. Be open to adjustment. It doesn't show weakness—quite the opposite, actually.
- *Communicate with your peers.* A common occurrence in many departments is the establishment of three different cultures on three different shifts. Of course, not all shifts will be run exactly the same. But a good way of keeping within arm's reach of each other is to talk frequently and discuss sore spots with other shift captains. What you don't want to happen is to have members transfer from another shift scratching their head trying to figure out how things run on yours. There is always an acclimation period when people switch up and go someplace new. But if the captains can get together periodically and compare notes on hot topics, this will help members to make a smoother transition.
- *Show your appreciation.* This should go without saying. Always praise jobs well done and convey that you are proud to lead your members because they make themselves easy to lead. If they aren't, then root out the cause and find the shift balance again. There is always an unseen cause to the effect you see.

Reflective Prompts

1. The shift captain is the rudder that can lead the team either to long-term success or down a dark path of chaos and uncertainty. Think about the people on your shift and jot down what you know about each one of them, from your lieutenants down. How much do you

truly know about your members? Are they married? What are their spouses' names? Do they have kids?
2. If you were to create a chart of strengths and weaknesses for each of your members, what would be in each column for each of your members?
3. Too often, strengths and weaknesses are discussed only during yearly evaluations. Do you know what each member is aspiring to accomplish? Are they working toward those goals? How are you helping them to achieve those goals?

9 The Chiefs: Where Culture Is Dictated

Leadership is solving problems. The day soldiers stop bringing you their problems is the day you have stopped leading them. They have either lost confidence that you can help or concluded you do not care. Either case is a failure of leadership.

—Colin Powell

Biting Tongues

Dean was the chief of the suburban department for a little over a year. He had tested well and beat out three other members to earn his place as the ninth full-time chief of his department, which had been around since the 1940s. Twenty years into his career and he had reached what some would regard as the pinnacle of the fire service for the town of 41,000 residents. There were 42 full-time members, and they ran about 5,500 calls per year.

Dean had many things he wanted to implement, but he had learned quickly that everything was a slow roll when it came to decisions made by city hall. His request for a new rescue squad was put on the back burner for another year. He had also tried to get money for personal thermal imaging cameras on all the vehicles. He thought they were becoming a vital part of personal protective equipment for structure fires and looked like a promising case to secure funding.

At this time, the mayor was in his second term in office, and Dean was a new kid on the block at department-head meetings. He tried to make things easier on his two assistant chiefs by making clear position duties for each of them, but those lines in the dirt were often smudged out. Overall, he still felt the transition was going smoothly, with the normal hurdles along the way.

Early in his time as chief, Dean noticed that he had changed his way of thinking after he put his name on the door and his badge on his shirt. He started feeling a sense of pride he had never felt before. Arriving at work early felt like his duty. Keeping his command vehicle clean was an obligation. His uniform was always pressed and clean, his office was impeccably neat, and his professional appearance was something he owed not only to himself but also to the public he served.

Dean believed wholeheartedly in the chain of command. From day one, he directed his officers that the members of the department would follow the chain of command to keep order and take a systematic approach to the daily routines and problem-solving. Dean emphasized this since there had been a few incidents when he was still a captain where the chain of command had been skipped, and high-level officers inundated with complaints from firefighters who had no business circumventing the process. Now, for the most part, the rules were being followed based on early returns from the shift captains.

As Dean walked through the kitchen toward his office one morning, he ran into Paul, one of the firefighters on shift.

"Morning, Chief," Paul said as Dean walked by.

"Good morning, Paul. How are things?" Dean replied.

"Not bad. Just starting out."

Dean noticed Paul's T-shirt was worn and had a faded blue to it. He didn't want to be a jerk, but he still felt compelled to mention it.

"What are you working on today?"

"I'm just trying to get a list together for pantry shopping," Paul said.

"Gotcha. Headed out soon?"

"Yeah, just about to leave actually."

"Hey, do me a favor and throw on a better shirt. Can't go out in public like that, right?" Dean said with a smile.

"Yeah, no problem," Paul replied. "Sorry about that."

"No problem," Dean said. "Thanks."

Dean walked up front and got his day rolling.

A few hours later, Dean walked into his office after a couple of meetings and noticed his garbage had not been emptied. At his department, garbage cans were to be emptied (or at least checked) on a daily basis by the crews in the morning during station cleaning duties. Just then, one of the members of the shift walked by in the hallway.

"Hey Curt, got a sec?" Dean called after him.

Curt backed up and stuck his head in the doorway.

"Hey, what's up, Chief?"

"When you get a chance, can I get this emptied?" Dean asked while he held the can up. "I have a council member stopping in, and that would be great."

"I got you. No problem," Curt replied.

Dean thanked him and Curt slipped away out of sight.

Both tasks were performed, and Dean thought nothing more about them as he went on with his business.

About 2 weeks later, the assistant chief, Jay, came in and sat down in Dean's office. They shot the breeze for a while and caught up on things like they normally did on a late Monday morning.

"What else is going on?" Dean asked casually.

"Interesting . . . I was talking to a couple of the officers on shift this morning. Did you mention anything about a shirt not being right last week or something?" Jay probed.

"I mentioned something to Redding before he went shopping. Why?"

"The crew was all up in arms at their next shift meeting, saying you were nit-picking about his uniform."

"Are you kidding?" Dean asked.

"No. I guess the guys were pretty blown out of shape about it," Jay answered after a moment.

"There was nothing to it," Dean said. "He looked like a mess, and I was completely nice about it."

"There was also something about you wanting your garbage emptied," Jay said.

"What!" Dean exclaimed, throwing his hands up.

"I'm just relaying the information," Jay answered. "Need me to do anything about it?"

"No thanks, Jay."

Jay got up and walked out of the office.

Dean was caught off guard. *What did I do wrong? I told those firefighters what I needed nicely, and now this?* Soon after that, Dean called the shift captain, Will, to his office for a chat.

Will and Dean had been friends for over 15 years. They had watched each other's careers blossom and had a very strong working relationship. They could speak candidly with each other, and Will rarely held back when he talked with Dean. Through it all, their interaction was appropriate for whatever stage they were on. Behind closed doors they usually spoke freely. That's why Will was there that day.

"What's up, Dino?" Will asked as he walked in.

"Grab a seat, Willy," Dean said with a smile.

Will sat down and looked around, casually. He then started the conversation. "I have a crazy feeling we're gonna talk about T-shirts and garbage cans."

"How'd you guess?" Dean said with a smile.

"I had a hunch. Yes, feelings were apparently hurt over you barking orders at Redding and Mitchell," Will said.

"Barking orders? What on earth? I didn't bark orders at anyone!" Dean responded, once again with his hands in the air.

"It's not what you said—it's what they heard," Will answered.

"So, I'm supposed to go around when I see something wrong and not say anything?"

"Not at all. It just depends on who you say it to."

"What do you mean?" Dean asked.

"If stuff is outta place, let me or the other shift captains know," Will said. "We know what's going on, and we'll handle it."

"Do you think it would have been handled if I hadn't said anything?" Dean asked.

"They already were handled before you said anything."

"How so?"

"Because Redding keeps a nice shirt in the bays for when he goes out in public, and Mitchell had a squad call this morning and hadn't gotten to garbage can duty yet."

Dean sat back in his chair. He was annoyed.

"And I take it I'm the jerk now for micromanaging?" he finally said.

"Pretty much," Will answered.

"How am I supposed to know what's going on with that kind of stuff?" Dean asked.

"You aren't," Will explained. "That's what I'm here for."

"Fair," Dean blurted.

"Hey, man, I got you. There's a ton on your plate all the time. Let us run our shift and come to me if something isn't right. Not the crew," Will said.

"Sorry about that, Willy."

"No worries," Will responded. "You have bigger fish to fry than shirts and garbage."

"Got that right," Dean said, suddenly feeling lighter.

"Anything else?" Will asked.

"Nah. Thanks, Willy."

"You got it."

Will stood up and walked out of the office, and Dean started shuffling paperwork, getting ready for his equipment budget meeting for the upcoming year.

Tools of the Trade

The scenario Dean dealt with wasn't the most earth-shattering, life-or-death decision that a chief may face. Certainly, other characters in this book have dealt with stickier situations so far. Still, what Dean's story lacks in style it makes up for in substance. It's not sexy, but it has claws. The situation was minute compared to other decisions a chief may make, but it is a real-life one that many chiefs come across—and many times can smolder under the radar before exploding.

Dean was meticulous about his presentation personally and clearly cared about the station and the firefighters. Unfortunately, while he banged the drum for the chain of command to be followed, he was the only person who truly broke it. The chain of command goes two ways, and by giving friendly reminders and making minor requests, he undermined his own initiatives as a leader of the department. Is it wrong to communicate with members of your department? Certainly not. But chiefs must be aware of the consequences of seemingly harmless remarks, which can get embellished and twisted down the line as stories unfold.

The assistant chiefs, captains, and lieutenants are tools for the chief to use. They all have their place—their managerial niches, to provide support. Delegation is the key to success and the only true pathway to the balance as a chief. As hard as it may be, chiefs must follow the chain of command they put in place. Have faith in your members: They went through rigorous testing and interviews to get to the position they are in, and they most definitely have a finger on the pulse of their shift.

Unless clear violations happen before your eyes or you are certain they are imminent, trust the process. If Dean had bitten his tongue instead of making the comments he thought were valid, both incidents would have been avoided, and nothing would have flared up. Instead, he indirectly created a situation that put added pressure on Will and his lieutenants to calm the crew down and get them back to center. Again, that's what they are there for. It was much more appropriate for Dean to have had a meeting with Will about perceived missteps he observed, rather than making additional missteps himself when he didn't have all the facts. As it turns out, there were no missteps after all, once Will explained each situation afterward.

The Balancing Act

The chiefs are not only in the ultimate position of power but also in a spot where virtually every decision they make is scrutinized endlessly. Different responsibilities will fall to any chief, depending on the size of their department.

In small volunteer and township departments, chiefs are saddled with the constraints of limited financial resources and personnel. Many times, the chief will be friends with members, and the hierarchy of authority and rank is scattered and broken if order isn't kept. This can cause organizational issues, and the reliance on response in rural areas puts an enormous strain on the person in charge. Regardless of whether justified, the chief will take on a sense of ownership that correlates to the successes or failures of a smaller department. If you find yourself in this predicament, it is essential to realize that there must be limitations to your department's expectations out of the gate.

We all wish for new apparatus, state-of-the-art tools, and terrific-looking stations that make us feel proud and our citizens feel safe. However, this will usually remain just that—wishful thinking—for long periods of time. There is only so much money to go around, and small rural departments have their hands tied financially. Don't put too much pressure on yourself and make sure your goals are realistic when considering the tools you have to work with.

In suburban departments, the chief works with other city departments and the mayor to plot budgets, stays within predetermined constraints, deals with meetings and public events, makes sure personnel and equipment issues are solved in a timely fashion, runs medium to large fire scenes, is the go-to for major events and tragic developments, keeps a finger on the pulse of the station through his or her officers, and still is a firefighter by nature.

A day in the life of a chief of a suburban department entails walking through administrative minefields. Few of the shift members have the faintest clue about what goes into this. Yet here the chief is trying to forge ahead and make the department better through choices that sometimes don't bear fruit until months— or even years—down the road.

Large metropolitan departments are a labyrinth of stations and personnel responding to every situation imaginable. The chief is usually tied up more prominently in the political arena at this level, and the people working directly underneath the chief become vital cogs in the wheel. It is not uncommon for midlevel firefighters to go a year or more without seeing their chief in person. These vast departments have been in existence for a very long time and work in particular ways to achieve consistent outcomes. Broken or not, procedures have been in place seemingly forever, and it takes an enormous amount of effort and cooperation to make changes throughout these Goliaths.

Most of the balancing between firefighter management and administrative duties takes place in smaller town and suburban departments. Besides sheer numbers, one main reason is logistical—namely, proximity. Few smaller departments have an off-site location for their chief to operate away from his people. Many times, the office is in the administrative wing of the central station. It becomes a very tight fit in a hurry. This tends to precipitate encounters throughout the day, leaving the door open to judgment and criticism. As hard as it may be, following the chain of command both ways is the key to effectively communicate, as we saw in Dean's example.

As a chief, you may think that some interactions are just "small stuff" and shouldn't be taken in a hostile manner. This couldn't be truer! You must realize that it is indeed just small stuff. Let it go. If it bothers you that much, then you need to say something, but say it to your captains or lieutenants; they will address your concerns with action to achieve what you want.

It seems trivial, but little things matter. In particular, they are not so little to the person on the other end of the exchange. It is hard enough to find the balance. Pick your battles wisely and ask whether this is truly worth a potential headache simply because something you saw irritated you for a fleeting second.

Transparency Equals Success

A common perception among the firefighters is that chiefs don't always do their fair share of work. The shift members come in all shapes and sizes, including their emotional makeup. Consequently, you could work your fingers to the bone around the clock, and it wouldn't matter to some of them. You'd still be perceived as a string-pulling puppet master who has been swallowed up by city hall, bending a knee to the evil empire in their view. Although a select few may deserve this, most chiefs have the well-being of their members as their top priority and the progression of the department as their objective.

Unfortunately, when allowed to draw their own conclusions, some members perceive an abandonment of camaraderie and a departure from their best interests by the chief when he takes his oath. They hope that the chief will do right and advocate for the boots on the ground, but most have their reservations. Think about it from their perspective: You got promoted to chief. You are no longer in the union. You no longer get up in the middle of the night for calls. You no longer have stories that weave us together and solidify a fire station's membership. You have passed over to the Dark Side. These are the myths you must absolve yourself of on day one if you are to have any chance of removing the fictitious barrier that has been put up between you and the members.

Several steps need to be taken to reconcile with the shift members. First, actually go to bat for your people. Many of the best chiefs were adamant about their stances in council meetings and mayor briefings. This isn't to advocate disrespect or abrasiveness in getting your point across to the powers that be. For the most part, chiefs want nothing more than to check all the boxes on their wish list with city hall on an annual basis. Who wouldn't? Often, it is easier to float the "all good" attitude regarding the needs of the department at meetings. But the old adage that the "squeaky wheel gets the grease" is true and extremely relevant to fire departments. Stop tabling things and get a commitment. You have an agenda of what your department needs in order to serve the public.

Second, verify with your members what the department needs are. In my department, we give a State of the Department address to each shift in January. This conveys to the shift members what the chief's vision is moving into the next year. The problem, though, is that the warm and fuzzy feeling that everyone gets from this lasts only until about February. The shifts are given an opportunity to voice their opinions and concerns. Shifts forget quickly what the endgame is because they like to see tangible results. If a new engine was green lit in a meeting in March, the shifts may not even know that the chief won this for them and it has been ordered unless it is spelled out for them as it happens.

The final step is transparency. Of course, you will not recite to the membership the minutes of every meeting you have with the mayor or go over every note you took during a council session. Still, if the shift members get a glimpse into just how difficult it is at times to get any movement on issues, they will appreciate your candor. Something as simple as letting them know that you spoke with the city council but they decided to stonewall you on some new equipment lets the members know that you're trying. A quarterly update is a great example of how to keep members in the fold so that they buy in to the department's forward movement together with you.

City meetings aside, another terrific way of connecting with the members is to give them a glimpse into a day in the life of a chief or assistant chief. Don't forget, while we have all been firefighters, only a couple of us have been promoted to chief. You know exactly what each member does on a daily basis because you've been there; the same cannot be said the other way around. The best way to crush a misconception is to hit it head-on by occasionally letting members know exactly what you do. Often, firefighters will greet you in the morning when they see you but then have absolutely no idea what constitutes a day for one of their chiefs. Meetings, presentations, budgetary issues, more meetings, problem-solving, updates from the shifts, more meetings—sheesh! Many firefighters would say "Thanks, but no thanks," if they had the slightest clue what you stuffed into each workday.

Remember, if you give these updates to the shifts, don't whine to them about how tough your life is. You're simply making them aware that you're only human, things take time, and there is a ton more you do daily than what they see. Soon they will realize that you deal with a myriad of issues when you arrive each day and that they are still a priority even though the signals get crossed occasionally. You'll find that the balance will be much more attainable when you have eliminated assumptions.

Roots

After you have achieved your goal of becoming chief, don't forget who you are and where you came from. You were once a cadet, a junior firefighter, a shift member, a junior officer, a shift captain (most likely), and maybe even an assistant chief before you ascended to your current position. The sense of pride you take in the accomplishments of your career is unparalleled. Some chiefs keep their old helmets or badges in their office to remind themselves of their firefighting progression through the years. It is a fulfilling and wonderful ride few members have the privilege to enjoy.

Instead of using those helmets and badges as office decor, remember what each one signifies. If you've been on the department for 20 years, you have gone through a ton of changes. From your psychological and emotional makeup to physical limitations that 2 decades has cast upon you, you've changed. Pause and recall what was important to you at different phases in your career. So many benchmarks were so important to you at each point. Think back to when you completed the academy as a cadet and walked across the stage for your certificate. Remember when you were a junior firefighter at your first actual ripper when you were on the tip. Recall your first time when you were in the right seat as an acting officer in charge because the shift needed you. Or your first time in charge on scene of a structure fire as a new lieutenant. Remember the day you sat behind your desk as a captain in the shift office and you realized you oversaw your very own shift. All these memorable milestones were important to you as you grew as a firefighter.

Now realize that everyone on your department is at various stages of their growth. They all have issues that they deal with that are important to them. Don't ever brush these off. There's a quirky caveat to the whole chain of command—again, it is a *balance*. Use the chain of command to convey concerns. Skip it when it comes to learning about your shift members. Use it in times of trouble and reprimand. Skip it in times of tranquility and praise. Do the delicate dance and put the effort in even when it seems it won't do much good; to the contrary, it works wonders. Just a minute here or there gets your rocks in a big enough pile that you have credibility and a record established with your members as someone who is reliable. Someone who is on their side. Someone who cares.

Final Thoughts

To review, keep the following in mind:

- *Problems are subjective.* The members on your department are all writing their own history. Remember, they are the main characters of their own story, and the challenges they face are important to them. Blanket reforms and sweeping policy changes may look good on paper, but do some market testing by asking your shift officers how these could impact their people *before* implementation. It is much easier to get a grasp on how things could go than to ask your officers to help with damage control after the gauntlet has been laid down.
- *Stay transparent.* Don't assume everyone else at the station knows what you are doing or gives you due credit for working hard. The shift members are usually going to look at you with a skeptical eye, and when they are left guessing exactly what it is you do with your day, they will fill in the blanks using their own imaginations. Medium-sized departments will deal with this situation the most; larger departments are simply too vast to realistically update all members on what is being done at the chief's level on a consistent basis; and rural and volunteer department members are usually in the loop with most of what goes on because of tighter relationships and smaller areas of concern to manage. If you are given the opportunity to keep your people aware of what is going on, do it. Even a small indication of what a day in your life entails will dispel rumors and correct the record.
- *Actions speak louder than words.* Get things done. Get movement on initiatives. A solid, detailed PowerPoint presentation doesn't mean anything if actions and results don't follow. On a chief's level of operations, you need to plan much more because of the far-reaching impact your decisions have on the entire department. Plan accordingly and then act. Like everyone else, a chief can make mistakes. Unfortunately, the ramifications and financial impact your mistakes have are much more severe than other missteps in the ranks. You can set your entire organization back if the planning doesn't match the execution. Finally, let your people know that part of the action you are perpetuating isn't always tangible. Progress is being

made, just not in a way that can be seen or felt immediately. They will appreciate your openness.
- *Don't take things personally.* You are in charge of a firehouse full of grown men and women who have opinions, problems, triumphs, personal agendas, and goals. If you have 50 people at your station, there will be 50 different stories and 50 different lives being lived. Part of our dynamic nature as human beings is that we have our own assessments of the world around us. As a leader of the department, it is not uncommon for you to take offense at opinions that essentially dismiss the hard work and dedication you have invested in your vision. Yet you must let things go. You won't please everyone, and taking things personally will drain your energy and shake your confidence as a decision maker. Lean on your ability to be a professional and keep business where it belongs.

Reflective Prompts

1. The department depends on the chief to make sure everyone is represented and the culture on the shift is one that cultivates growth and development for all members. Do you know what the goal is for your members this year? You needn't know this off the top of your head, but you should be able to picture each member and recall one thing they want to accomplish for themselves.
2. Do you check in periodically with your officers to see what the state of the shifts are? Has a State of the Department address ever been given to you?
3. Going a step further, talk with your captains to see what they have come up with for each individual on their shifts. This is an excellent way to keep your finger on the pulse of personal stories and growth.

10 Two Homes

Work, love, and play are the great balance wheels of man's being.

—Orison Swett Marden

Oops

Kevin rode home from the station the way he always did in the summertime. He checked the weather the day before, and with a clear forecast he rode his motorcycle in, taking the long way. His route was usually quiet when he zipped in on the weekends and he was the king of the road for 15 minutes or so. At his wife's beckoning, he reluctantly wore a helmet. The trip home was usually more hectic during the week because of traffic, but he still enjoyed his ride. For 12 years he had been on the department, and although he came up short two times on the promotional examination, he was still a good firefighter and took pride in his work.

Trina and Kevin had been married for 6 years. Trina was a first-grade teacher, and they had met after being set up by a coworker of hers. They had a 3-year-old named Georgia, and there had been discussions about adding a second rug rat to the household soon. They parented well together and were still connected in a lot of ways. They planned date nights and attended functions together but still had space from each other devoted to their respective friends and families. Overall, things were going well, and there was a rhythm to their life together.

To unwind, Kevin would usually go to the gym before arriving home to take over parenting duties with Georgia when Trina left for work. Then he would have his little helper by his side while doing yard work and fixing things around the house in the morning. After lunch and an afternoon nap, dad and daughter would head to the park for a bicycle ride and then a quick stop at the playground before stopping for ice cream cones on the way home.

Summers were awesome. Trina was off school, so Georgia was with mommy all the time. This gave Kevin an opportunity to join his buddies in a round of golf here and there on the weekends. Trina also enjoyed hanging out with her girlfriends now and then, hitting happy hour for drinks and venturing out to a winery occasionally.

Their lives were clipping along nicely until about 3 weeks prior to this particular day. They had begun to bicker. First this was over little things. But soon the fights become bigger ones. Trina was the communicator in the marriage. She liked to get things out in the open, and she excelled at this. Kevin tried his best to reveal his emotions but failed as often as he succeeded. Trina felt as though the relationship felt "stuffed up" over the past few weeks. She and her husband still talked, but conversations had downshifted to surface material lately.

So, when Kevin came home one warm summer morning after his shift, the powder keg was ready. It needed only a match to light it up.

Kevin came in the door while Trina was at the table, sipping her coffee and looking out the back bay window. Georgia was in the other room playing, chattering to herself and trotting her toy horses around the floor with cartoons on the television in the background.

Kevin strolled by Trina and walked to the cupboard for a cup.

"Hey," he muttered.

"How was shift?"

"Fine. Busy."

There was no response from Trina as she flipped through her phone, set it back down, and stared out the window again. Nothing further was said until Kevin poured his coffee and sat down reluctantly at the table. Trina shot him a look and he got up and walked in the other room to kiss his daughter on the top of the head while she played. He plopped back down at the table.

"How is everything around here?" Kevin asked.

"Your sister still wants to go shopping later?"

"I don't know. She hasn't called yet." Kevin paused before continuing. "You OK? Did something happen?"

"Same as usual. Oh, Gigi learned a new word yesterday."

"Oh yeah?" Kevin asked. "What was that?"

Trina paused, thinking. Then she uttered, "Um, let's see, she was playing with her ponies, came into the kitchen, and asked where the hell her juice was."

Kevin smirked, then giggled. "She knew what she wanted."

"You think that's funny?" Trina questioned, irritated.

"C'mon, T. It's what kids do."

"Not our kid. She picks up on things, you know that."

Kevin slammed his weight back into the chair and folded his arms defensively. He resembled a scolded 8-year-old.

"So, this is my fault?" Kevin shot back, "You've got to be kidding me!"

"No, I'm not," Trina said, still holding it together. Her upper lip curled as her own frustration was itching to come out and play.

Kevin got up from the table and poured his coffee into the sink. He was fuming now.

"It's one word!" he blurted out, now shouting.

"Easy. I'm trying to talk to you, and I'm obviously upset," Trina said. "And your daughter is in the other room."

Kevin angrily pulled his shirt off, slung it over his shoulder, and stormed upstairs. He had had enough of the accusations. Trina had stood from her chair at the table but slowly sat back down. Gigi came into the room and sat on her lap with her favorite toy horse, Prince, in her hand. The little girl could sense something was wrong. Kids and pets are funny that way. Both are extremely intuitive about stress. Trina kissed her on the head and played with her daughter's hair.

Kevin did his best to keep busy that day, working outside and fiddling with the muffler on his motorbike. He went as far as eating a protein bar he found in the garage instead of chancing a run-in with Trina in the house while rummaging for food. As the afternoon wore on, he looked for other things to clean and organize that really didn't need it, putting off resumption of the conflict—as well as its potential resolution. We've all done this.

Trina came out to the backyard around 2:00 p.m. Georgia was down for her nap, so she thought it would be a good time to discuss things. *What was his problem? Why was he acting so abrasively?* She found him by the back fence organizing the shed. Equipment was scattered about, waiting to be placed neatly back inside. She had brought a sandwich and chips for him, along with a Gatorade. He saw her coming and wiped sweat off his forehead before going inside the shed to remove three snow shovels, one of which had a broken handle.

"Hey, got a minute?" Trina asked as she approached. "Gigi's sleeping."

Kevin paused and looked at her, nodding slightly but not saying anything.

"Kev, I don't know what's happened, but something's been up with you for a while now."

Kevin avoided eye contact and moved a couple of boxes of outside Christmas lights.

"I mean, you've been short with me and your daughter, and I don't know what's happened," Trina said.

"Just stuff going on," he finally answered.

"Is it your mom? I know she's been needy lately."

"No," he said abruptly.

"Well whatever it is, our daughter is picking up on it, and the language thing is just the start."

"C'mon, she hears that 20 times at family functions!" Kevin fired back.

"I've noticed you've been swearing a ton lately, and I'm trying to figure out why," Trina answered.

"I'm not gonna keep score over how many times I swear."

"But *why* are you doing it?"

"I don't have to answer to you on why I swear. I can't stand when you act like you're my mother when we discuss this stuff."

"You're not answering me at all. And can't I ask my own husband why he's acting strangely lately?"

"She died!" Kevin blurted, emotional and embarrassed.

Trina was taken aback by the admission. Kevin sat on an old lawn chair and rubbed his face with his hands. Trina walked over next to him and put her hand on his shoulder.

"Died? Who?" Trina asked, the furrowed brow disappearing.

"The 4-year-old from the motor vehicle accident died," he said.

Trina paused. The snarl had completely left her face. Her eyes softened and her blood stopped boiling.

"When?" she finally managed.

"Two weeks ago Monday. She was in the ICU [intensive care unit] a week before that."

"Oh no. I thought she was doing better."

"She was, but she took a turn I guess. I couldn't get all the details."

"I'm so sorry."

"Sucks," Kevin replied. It was all he could muster.

"You did all you could," Trina said compassionately.

Kevin stood up and they embraced. There was appropriate silence for a minute. While they hugged, he calmly apologized in her ear.

"Sorry," he whispered. "This is a tough one."

He looked at her and saw how much she cared and was hurting for him.

"All I could see was Gigi throughout the whole thing."

"I'm sure," Trina said.

"I'm sorry about the last few days. I didn't want to trouble you with it too."

"I'm on your team," Trina said. "Remember that."

As the two of them hugged it out again, Kevin felt the stress releasing from his body. It felt good to let it out finally. He gathered himself and felt a little better. He glanced at the food she had brought him on the small table nearby, took a bite of the sandwich, and washed it down with the Gatorade. He wiped off a couple lawn chairs that were nearby, and together, Trina and Kevin sat down and caught up finally, with renewed interest in each other's lives. Eventually the waters in his head receded, and he was able to level off again emotionally.

The Pathway to Success

We all have a home life. This may or may not be synonymous with a spouse, kids, house, and so on. Regardless of whether we come home to everything I just mentioned or live with pets or alone inside our own little bubble, we have a distinct separation between what we do at the station and what we do at home. It is essential that we keep in mind that there will be times when one life bleeds into the other. This can be a two-way street.

Our job calls for rapid response to situations from which many people in society would withdraw, cringing. We deal with almost anything—from burn victims, full arrests, and violent motor vehicle accidents to dying patients with their families pleading for us to help them and even pets that have been lost. All of these are traumatic events that most people dread, yet there we are, seemingly impenetrable, stepping up and handling the call. But we're still human. We still have feelings deep inside that cause us to take stock in who we are and what we do.

Often, we take a different route with our emotions. Other people may break down and show extreme reactions and sadness after witnessing a traumatic event, even one that is mild compared with what we are usually subjected to. But we are programmed differently: We toughen up usually by compartmentalizing or burying our emotions to keep ourselves from being vulnerable. That can work, but we have to stay on our guard when we leave the station. This is because the pain we feel often manifests in other ways, consciously or subconsciously, as we deal with our own emotional trauma.

In the story, Kevin was becoming irritated at home and began to lash out in passive-aggressive ways at his wife because of the child lost on a recent call. Luckily, his wife was dialed into his emotions and identified that he wasn't behaving like himself. Trina was able to confront his emotional turmoil in a nonthreatening way, and Kevin then disclosed the source of his irritability. Unfortunately, many firefighters are programmed to bury those emotions, and the friction that results in their lives can go on for months and even years because of their determination not to look vulnerable. This is extremely common—unhealthy, but still common.

On the other side, home stress can take a powerful toll on us at the station. We have bills to pay, families to raise, relationships to maintain, parents to deal with, and the list goes on and on. Almost everyone in the private sector knows where we are coming from in that regard. The difference for us is that we are then asked to deal with the tasks I described earlier with a clear head and a sharp mind. If there is any compromise to our decision-making faculties, bad things happen to us, our crew, our customers, and anyone else involved. A clear mind is paramount to our success.

Unfortunately, as already stated, we are only human. We may come to work riled up by a fight with a spouse who is on our nerves or a kid who wrecked our car and start the shift of an extremely stressful job with . . . stress. How can we find the balance then? For starters, communicate. Without divulging too much personal information, you can always let your brothers and sisters at the department know that you're going through a rough time at home. Even a minimal insight can go a long way because it tells the people you work with that you're sorting through things, and they will give you some leeway.

Anxiety at home is typically easier to mask when we are at work. This is usually because of the environment we find ourselves in at the station. The work family tends to gravitate more to sarcasm and cutting remarks much more than our family at home. Thus, turbulence at home may go unnoticed for a very long time at the firehouse. We can get lost in the shuffle, and nobody notices. That's why we all have an "I never saw it coming" story about a divorce of one of our shift mates. One day they just show up and blurt out that they are ending their marriage, and this turns everyone on their ear. The writing had been on the wall for months or longer. We simply never saw it—and usually we were never invited to.

Whichever way our stress is flowing, the gateway to success is (gulp) transparency. The word is thrown around rather casually. In fact, if we are emotionally transparent with our lives, we are able to alert those around us in either family when we are out of balance and give the reason why. Suddenly, we have allies instead of adversaries. People we perceive in our lives as obstacles (because we are hiding emotional hurdles) become understanding when they know what's eating us. They may avoid a possible confrontation over something small knowing you've been strung out recently over a valid life situation.

Let's be clear, though (pun intended). Transparency of this kind does not mean you come to work or arrive at home every day with some new problem bothering you. I'm referring to major events that knocked an otherwise solid pillar of the family or the shift off his feet. Let those around you know what's going on. Set your ego aside (more on that in chapter 13). However you need to convey it, get it out. If you don't, it will grow much larger in one way or another.

Manifestations

Negative developments in our lives affects us in numerous capacities. We can physically feel our body shift in response to new stress levels. Our emotions turn us into something completely contrary to how we normally behave and interact with others. Matters that typically would not get a second thought from us become inflamed and magnified. Some of it we can catch. A lot of it we can't. So how does our stress typically manifest itself on the two home fronts?

Let's start at the station. We are creatures of habit, as we all know. We plot our days off, hope for light call loads, pray to get at least four to five hours of sleep overnight so we can function the next day, and basically try to do the best we can when we punch in for our shift. When things go sour at home, we tend to lose the balance. We start to go down a path of deterioration that follows a typical pattern.

The first feature of this pattern is that we lose our interpersonal skills. A member going through difficult times will typically start to circle the wagons emotionally and begin to shut down in their own way. They tend to become irritable and generally out of character. Some might become quieter than usual, while others respond in an equally detrimental way by lashing out at fellow members and getting mean, using personal and cutting remarks. Shift members can usually pick up on this, but again, it takes time.

A second characteristic of this pattern is a general irritation with otherwise mundane or normal duties around the station. Out of nowhere, complaining becomes a theme, for example, over such matters as cleaning, meetings, continuing education, or training. This is typically because when we are going through emotional turmoil, we want to have some sort of hiatus from other responsibilities to give us time to sort through our thoughts and to decompress. Remember, when we have problems at home, work becomes a sanctuary of sorts. How dare we have to work when all we want to do is take a breather from all our problems at home?

The third and most detrimental casualty of home stress is when our work suffers. Red flags are raised when we begin showing up late for work and missing essential skills we need to perform. When our head isn't in the game, people can get hurt or even die. Our focus must stay sharp and true, or else we are going to push the wrong medication or forget our firefighting techniques. That could end extremely badly for us or someone else.

When we are having a rough time at the station, stress can manifest at home in a plethora of ways. Depending on our dynamic with our significant other or our family members, we begin to act out of character. There are too many different relationship situations to list here, but regardless, things truly seem . . . out of balance.

This stress may manifest as a lack of interest in hobbies, a shorter fuse with the kids or spouse, a desire to spend more time alone, and a general insulation from close friends and extended family. Most of the time, this insulation will take place because the member is not yet ready to discuss what the cause of their emotional stress is. When you subject yourself to people closest to you, they can usually read you like a book and can pick up on the small nuances you reveal. Before you know it, you're exhausted from trying to put on the act of normalcy, and the beans are promptly spilled.

Let this serve as a heads-up not only for yourself but also for those close to you at the station. Read the signs of someone who is hurting and working through problems. You can be a resource to them as they navigate a very personal and tough situation, whatever it may be.

Finding the Balance Between

The key to finding the balance between the two homes you exist in comes down to five simple areas: communication, compartmentalization, self-care, awareness, and consistency.

Communication

Throughout this chapter, I hope you have recognized that communication is incredibly important to finding the balance between your two homes. If you already have good communication skills and have been able to convey both good and bad situations on both fronts, you are well ahead of the game.

It is absolutely critical to keep the lines open. It is much easier to tell funny stories and triumphant tales when there is a good outcome. The challenge arises when you have to look vulnerable because some sort of nastiness reared its head and you have been knocked off center. Suddenly the focus of the story is no longer on someone else and *their* situation. You are now the center of attention, and all eyes are on you and how you dealt with or are trying to deal with your dilemma. That can make life very uncomfortable for us.

Although men and women convey their emotions in different ways, they both need to find a way to bridge the necessary gaps to keep each other familiar with situations at any given time. Firefighters in general have a hard time with certain words when they hear them. Being "emotional" has taken on a negative connotation over the years when in fact it is an excellent quality to possess. We are programmed to simply bottle up their emotions.

Regardless of what sex you are, communication either at work or at home is paramount to avoid mixed signals, dissolve hidden resentment, and above all make life much smoother as you process problems.

Compartmentalization

This is tricky but essential. When we compartmentalize our problems either at work or at home, it can usually get us ready for the environment we are heading into. Healthy compartmentalizing can be good.

As we roll into work, we put our game faces on and focus on what awaits us. We know that calls will pop up, training might take place, and small problems will inevitably require our attention.

After our shift is over and we head home, we must shift our attention to what responsibilities lie ahead as moms, dads, friends, neighbors, and all the other roles we play. We forget much of what we dealt with at work and must now refocus once again on our other life.

When we compartmentalize, it feeds into the old adage of "Leave your problems at the door." This can be invaluable, and we do it automatically about 90% of the time; the other 10% is the part we have to deconstruct and analyze to see if the compartmentalizing is healthy or not.

From a healthy perspective, compartmentalizing occurs when we arrive at work and take whatever problems we may have at home and shelve them, to work with the cleanest slate possible. This allows us to perform at our best and find our balance for that day.

What we must guard against is the unhealthy side of compartmentalizing. This is when we have an incredibly difficult time at work or at home and suppress emotions in an effort to create the illusion that we're fine. Sooner or later, the lid blows off: The emotion we have been suppressing because of our mountains of troubles explodes all at once, like a volcano.

Sadly, one of the great casualties of our profession is a complete collapse of emotion. Instead of processing them, we bury emotions into a little corner of our mind for long stretches, and we become even worse: We become *emotionless*. Nothing will drive a relationship to the brink of collapse quicker than an emotionally bankrupt partner in your life. Be aware of this trap and stay the course. The balance will follow.

Self-Care

This is a monster that many of us overlook. (I am guilty of not practicing this at times and have to keep myself in check.) It is extremely easy to fall into the trap of taking every stitch of overtime, joining every special team that comes our way, and exhausting ourselves by working other jobs on our days off for the sake of getting ahead. Weeks turn into months turn into years. We become human *doings* instead of human *beings*.

Stop and take a break once in a while. It's OK. If you take a nap during the day, don't catalogue it in your head as not doing anything. Napping is what you are doing at that particular time. Taking time and giving your body, mind, and spirit a break is not only advised, it's essential. Pursue your hobbies, integrate into your family, and make sure you allow yourself to have fun. You dictate how happy you are, and don't rely on a pile of money as the endgame to deter you from what's really important in your life: family, friends, relationships, and life.

Awareness

What it means to be aware of things in your life could take up an entire book by itself. The practice of awareness is a potent one that enables you to watch yourself and be a gatekeeper in decisions you make to find and keep the balance. When you practice awareness, you are well on your journey toward the biggest prize of all: finding the complete balance in life. The good thing is that this is achievable by anyone. The bad thing is that people from all walks of life have tried throughout history to harness this unparalleled power and rarely achieve it fully.

This is not meant to be a deterrent to trying. Even a small amount of awareness in your life puts you *way* ahead of the masses. But what exactly is it?

Awareness is the ability to recognize your surroundings objectively and nonjudgmentally. When used effectively, it allows us to calm ourselves down and make rational decisions, regardless of the situation, without the cloudiness of emotions getting in the way. Many people pursue this fundamental skill through yoga and spirituality. We can do it daily. With practice, you can learn to calm your mind and train it to be a useful tool, not an obsessive opponent to the balance.

Awareness also plays a role by keeping us attuned to the other people in our lives more fully. We develop a natural empathy for our loved ones at home and our brothers and sisters at the station. Things flow much easier when we are aware that other people have a story, too, that they bring with them whenever we are around them.

Consistency

The final link to successfully connect your two homes is consistency. When you are consistent in your relationships and with yourself, you can identify when you become out of balance much quicker than you could if you were constantly unpredictable with your behavior and thinking patterns.

There are a series of checks and balances that work beautifully when the people we care about at work and at home know what they are getting every time they see us. They know our tendencies, our mannerisms, and our overall well-being. When things get knocked off center, we are assisted by these people, and they help us to decipher exactly what it is that is derailing our normal routine. Be consistent with those close to you. You'll find that the balance is much easier to obtain if they know what they are getting when the best version of you shows up consistently.

Your Houses, Your Homes

Practicing the elements I just mentioned can assist you in finding a balance between work and home. Without some type of framework, we are left loosely meandering through life until negative tendencies develop that we really don't want; moreover, these tendencies become habits that are difficult to break.

The suggestions in this chapter hopefully will be a springboard for your personal development as you establish your own parameters for a healthy, smart balance of your life. Take control early and set down the pathway to your success. The more proactive you are with practicing your own elements to live by, the more success will come your way and the quicker your balance will be found in this most critical of areas.

Reflective Prompts

1. What are five key elements that are important for you to find and keep the balance between your two homes? Are there areas where you feel you can do a better job?
2. Make a list of what you'd like to improve on. Be honest with yourself and don't get lazy: There is always something. It is extremely important that you recognize your strengths and weaknesses. It is much easier to address a weakness you have found in yourself during quiet reflection than respond to it as an incredible revelation someone else eventually sees that you missed.
3. Is there one aspect you believe is the key to your balance? Make it your mantra. And then live it.

11 Dealing in a Healthy Way

In three words I can sum up everything I've learned about life: It goes on.

—Robert Frost

A Close Call

"One more round on my tab and then close me out," Jake said as he took a swig of beer.

The bartender nodded and put in the order. Jake went back to talking to the two women he was sitting next to at the bar. He was playing the game, but so were they. Both women had been laughing at mildly amusing jokes to pass the time and to milk his wallet for the better part of 2 hours. They were showing just enough interest to keep him on the hook. He was getting sloppy, and they were getting bored.

"So . . . Oh yeah, you were talking about Paris," Jake said.

The woman sitting closer to him looked incredulously at her friend and rolled her eyes. "Yeah, about an hour ago," she uttered while scanning her phone and snickering.

Her smile was long gone. He had had a shot in the beginning. His build and his eyes got him in the door. But as the night flew past midnight, Jake was all over the place with his questions and going nowhere with his stories.

He had been on a large city department for more than 6 years. Single, he was convinced that building a strong foundation of wealth on the department and getting a head start on his nest egg from his roofing business on the side were more important than settling down with someone.

From a firefighting perspective, Jake was near the bottom of his class in the academy. Yet he did well in interviews and excelled during the agility test.

Jake vaulted up the rankings, and one of the battalion chiefs took a liking to him when he said his father had been a U.S. Marine. When the city finally gave the green light for hires, Jake was one of 44 cadets who accepted a position and became part of the family.

His career was going fine. He survived as the new member and took his lumps from the mild verbal abuse that comes with it. But he liked his job. He didn't really cement his place at Station 22 until he pulled a badly burned child out of a second-story window during an apartment fire 2 years prior.

Jake was always on the periphery of friend groups growing up, and while he made progress socially in the academy, he never completely made consistent friends. His only true connection with coworkers in the department socially was hitting the bars on days off and hanging around the clubs when the 9-to-5 crowd had free time on the weekends.

Soon that well dried up too, and Jake decided to plug himself into the club scene downtown. He was a social butterfly and because he did decently with the roofing business, he had money to burn.

Over the past year, he decided the roofing business was taking away from his personal time, so he backed off the side gig for a while to enjoy his 20s. The station became more of a hobby to support his social life, and although he was still adequate at his job, his performance was sliding.

Jake was sternly reprimanded by his shift captain for forgetting the irons during a door breach on the south side of town. He forgot the first-responder bag when he was repeatedly reminded not to during a full arrest a month ago. He slept through three overnight calls in the past week alone. Things were starting to get dicey for the 26-year-old.

The women at the bar were done with Jake. The bartender returned with their drinks and set them on the bar. Before Jake could even sign his name on the tab, the two women had disappeared into the crowd, making their way to stairs leading to the second-floor area technically reserved for very important (or attractive) persons.

Jake shrugged and took a sip of his beer as he sat there alone. He tried to look attentive and approachable, but everyone knows how that story ends. At 2:15 a.m., he headed for the parking lot.

Jake found his pickup truck and headed home. He lived less than a mile from the bars, which came in handy most nights. He made it home OK, threw his keys somewhere near the kitchen counter, and fell soundly asleep in his bed with his clothes still on. It wasn't until just before 8 a.m. the next morning that he regained consciousness. Jake did a double take on seeing the time, snapped awake, and made a quick call to the station.

Once dispatch answered, Jack spoke in his best "sick" voice, "Hey, it's Bradshaw from Station 22."

"What's up, Jake?"

"Calling in sick for the first 12. Can you let Phillips know?"

"Will do," dispatch barked back.

Jake laid back down and took a deep breath. That was a close one. It wasn't until after noon that he woke back up. He jumped in the shower and watched a cable movie marathon until around 5:00 p.m., when he decided to call off the rest of the shift. He called a couple firefighters from Bravo Shift who were busy, but he still got ready to go out. No sense in wasting a Saturday. This was beginning to be Jake's unfortunate routine.

About a month later, Jake showed up for work in the nick of time and punched in a whisker's breadth from being tardy. He could tell something was off that morning. He was sweaty, pale, and just felt crappy. The morning went without a hitch, but by the third sweat-soaked T-shirt change before noon, the shift was beginning to notice.

Jake had to go to the bathroom for about the 12th time that day when a structure fire tone rang out. He ran out the door, into the bays, donned his bunkers, and hopped on the truck. It took everything in his power not to pass out in the process. Engine 22 steamed its way toward a fire on the lower east side.

The other two firefighters in the back noticed Jake struggling as he was putting on his self-contained breathing apparatus straps during the response. One of them felt compelled to say something.

"Hey Jakey, you OK?" asked Owen, who was on the irons.

"I'm fine," Jake said.

The other two looked at each other and shrugged. Three minutes later, Engine 22 pulled in front of a two-story duplex that was showing smoke from the window of an end unit on the Bravo side, first floor. Jake's shift mates jumped out of the engine, pulled a line, and were going on air when Jake stumbled into the front yard and had trouble bending over to flake the line out for them. *How the hell were they moving so fast?*

Jake kicked at a kink in the hose, lost his footing, and stumbled to one knee. He tried to pull the hose straight, but his hands were shaking. Wiping the sweat from his brow, Jake tried to steady himself enough to get back to his feet.

The lieutenant on the engine, Rob, had seen enough. He had just finished his 360° evaluation and knew that Jake was toast.

As Engine 30 arrived on scene, Rob radioed in. "Eighth Street command to dispatch," Rob said coolly.

"Dispatch. Go ahead, command."

"360 complete. We have a working house fire on division one of the structure. It seems to be coming from a bedroom on the first floor. We will continue in the offensive strategy with suppression and primary search."

"Dispatch copies: continuing in the offensive strategy."

"Command to Engine 30, bring your crew up and grab a second cross lay to assist the attack from Alpha."

"Engine 30 copies: pulling a cross lay to assist."

Jake, meanwhile, had gained his footing again and was trudging toward the front door. Then a firm hand grabbed his arm from behind and almost flung him to the ground.

"Go back to the engine," Rob instructed severely.

"I'm fine," Jake said in a last-ditch effort to keep it together.

"Go back to the engine and take your bunkers off—now!"

Jake reluctantly turned around and walked back to Engine 22, defeated. Engine 30's crew blurred past him with the cross lay and began their assault with the rest of Jake's crew. Jake sat on the bumper of the engine and began to dry-heave at his own feet.

Enough crews got there to knock the fire down and the duplex was evacuated. Only one person was found in the other occupancy, and she was unharmed for the most part. There were no injuries and the exposures had been protected. An emergency medical services crew on scene ran a quick check of vitals on Jake. He was tachycardic but otherwise OK.

When the crews got back to the barn, Jake was ordered to go home to deal with his health issues from the day. He could tell that Rob was fuming over the incident, but hey, stuff happens and he didn't feel good.

Jake got home, showered, and relaxed. His sister called around 10:30 p.m., but he didn't pick up. He poured a couple shots of vodka and unwound on the couch. Two hours and half the bottle later, he felt fine and rejuvenated. It was a Tuesday, so he decided to stay in for the night.

On Friday, the alarm on Jake's phone sounded, and he got up slowly out of bed. It was his shift day, and he just wasn't feeling it again. He called dispatch to take the weekend to get himself straight. He told himself that come Monday there would be a brand-new Jake at the firehouse.

The weekend went by in a flash like it usually does. Three clubs and a sleepover at a new female friend's house constituted Jake's festivities.

Monday is always waiting, and the clock rang with its 5:30 a.m. alarm chirping. Jake got out of bed feeling weak and exhausted. Maybe it was the flu. In any event, he had to get to work to discuss the events of the previous week's fire. He gathered himself together and noticed his hands shaking ever so slightly as he showered and brushed his teeth.

As he walked through the kitchen, he noticed a plastic bottle of vodka sitting on the counter with the cap off. *To drink or not to drink . . .*

Jake stared at the bottle. Every ounce of reasoning told him to grab his keys and walk out the door. But he couldn't. Not today. Just a swig to center himself and some gum in the car would make him a new man, though not the same one he'd promised. He skipped the glass. A swig turned into a gulp turned into a guzzle. He wiped his mouth and left in a hurry, the door of his apartment slamming behind him as left for the station.

It took about 20 minutes until one of the older members at the station, Alex, noticed an alcohol smell when Jake walked by in the kitchen. Alex picked up on it right away and appropriately notified Rob. It wasn't long before Jake was sitting in the shift office across from his lieutenant.

"Jake, we gotta talk," Rob said bluntly.

Jake nodded quietly and glanced at Rob before staring down at his feet.

"You've done some good things your first few years here, but what's going on?" Rob asked inquisitively.

"I don't know, just haven't been feeling good lately."

"Well you seem better today," Rob answered.

"I do feel better, actually."

"Couple guys smelled some booze on you this morning. You been drinking?"

"What? No. I mean . . . who told you that?" Jake asked, irritated.

"Not gonna tell you that. Have you or haven't you?"

Rob swiveled in his chair and put on his readers. He opened a folder and adjusted himself in his chair. "Before you answer, listen very carefully," Rob said.

Jake threw his weight forward in his chair, placed his elbows on his knees, and stared at the floor between his feet. His defenses were up.

"Have you seen your attendance over the last 3 months?" Rob asked.

"Called off a couple shifts like everyone else."

"More than a couple. Do you realize you've called off eight shifts?"

"No way," Jake said. "Not possible."

"We can check the timecards if you want. But it's right in front of me."

Rob closed the file folder and took his glasses off. He folded his hands in front of his face and began again: "As far as the fire went, I know you were frustrated with my decision. But I didn't see you fit to go in there."

"I would have been OK," Jake responded.

"Sorry. I made my call. You were having trouble walking. Not gonna send you in that way."

Jake leaned back slowly in his chair and eventually settled his gaze on Rob.

"So, I'll ask you again if you've been drinking today. Because you're right, you look a lot better today than you have in a while. And it makes sense if you've been drinking."

"What are you implying?" Jake said, raising his voice.

"I'm stating facts right now," Rob answered calmly.

"Well, they sound like accusations to me," Jake fired off.

"One way to find out if you want to go that route," Rob said. "One phone call and we can clear this up for sure."

Jake slumped back in his chair, defeated.

"Oh man!" Jake yelled, standing up quickly. He started pacing around the room, nervously.

"But I'm not gonna do that," Rob said as he followed Jake with his eyes.

"So what, I'm just fired?" Jake said, now yelling.

"Never said that. I'm on your team. Let's work through this together."

Jake sat down again, holding his head in his hands.

"Have you been drinking today?"

A weak nod without picking his head up.

"OK, we're making progress. I'll send you home today again, and I'll talk to the chief and let him know what's going on."

Jake slowly revealed a red face full of anger, fear, and embarrassment as he refocused his gaze on Rob.

"We'll get a plan together to get you some help. We've got this."

Jake got up slowly and took a deep breath.

"I'm sorry, Rob."

"Don't be. Proud of you, kid," Rob said. "Get your stuff and head on home."

Jake ended up doing well. He entered the employee assistance program and completed the necessary steps to return to duty without major repercussions. Rob handled a tough situation in an exemplary manner. I will touch more on him later.

On Our Own Path

Stress is a killer. Numerous statistics and data have confirmed that stress can take an enormous toll on the human body in several ways. The direct pressure we feel and exert on ourselves can lead to immune deficiencies, cardiac complications, and even the acceleration of debilitating diseases such as cancer.[1]

These symptoms and effects are felt even more profoundly in the fire service. The amount of stress we deal with triggers varying adverse reactions, including posttraumatic stress disorder, depression, alcoholism, and suicidal ideation.[2] This is not about any of those direct impacts. We need to focus on how we react to the stress in our lives, given the daily pitfalls we must navigate.

As our careers advance and we become acclimated to what we routinely experience, the stress will usually originate from one of two sources. The first source is the atypical call we go on that sticks with us or shakes us up. Such a call involves one or more of our senses and strikes a nerve—in other words, it is like an inflamed reaction in our psyche. The emotional drama caused will stir something inside us that we can't handle, and the call becomes ingrained as a terrible memory. Each of our senses can be involved in these:

- *Sight.* We see horrific events, and images of these may be burned into our heads, to the point that we replay the mental image over and over. Burn victims, gunshot wounds, inconsolable family members clutching loved ones, dead or dying pediatric patients, a fellow firefighter falling in the line of duty, just to name a few. That's heavy stuff. As mentioned in previous chapters, we are still human. There is only so much we can do to insulate ourselves from the trauma inflicted on our emotions.
- *Sound.* We can still hear the screams many years after they are gone. Cries for help. The gut-wrenching wailing from someone in great pain and suffering. The audio is logged into the recesses of our minds and is incredibly hard to cope with at times.
- *Smell.* The stench of death is something we have all dealt with at some point in our career. It is an odor that can't be mistaken for anything else once it has entered your nose the first time. Other smells may also be a trigger for someone when associated with a particularly difficult call. For example, a specific perfume that someone may have worn during a cardiac arrest or something as simple as an air freshener smelled during an extrication of someone killed in a motor vehicle accident. Among the five senses, smells bring the most vivid memories to humans in recalling details of an event or person.[3]
- *Touch.* I can still feel the rib cartilage cracking under my hands during the first compressions I ever did on a geriatric patient. The input we get from the sense of touch can stick with us for a very long time. The feeling of intense heat during interior operations is unmistakable. Touch can be a powerful recall tool of many situations we have encountered in our careers.
- *Taste.* The taste of soot and other residue after a fire can be powerful. Inhaling at the wrong time with unlucky wind directions can leave a taste of smoke and combustion that will make the most hardened firefighters gag. Taste is not as prevalent in these memories as some of the other senses, but it can still leave a lasting impression in our minds.

Whatever the trigger for our recall of an emotionally traumatic event, a door is opened to a memory that has struck a nerve with us, and it will linger until it is addressed.

The second source of our stress comprises the culmination of several small events that eventually wear us down to the point that we must deal with them as a collective enemy. There are not only events we run into on the job but also situations we find ourselves in. Regardless of whether the circumstances are of our own doing, in these events or situations, we feel as though the walls are closing in and desperately seek a way out. Whether at work or at home, some type of mental upheaval has occurred. Essentially, we have arrived at a crossroads in our head that has signaled to us that it is time to deal with the problem in one way or another.

These sources can also be a combination of several different stressors, including failed relationships, dead or dying relatives, a string of calls that have worn us down, an unhappy work environment, or simply that we are not where we envisioned ourselves at a certain stage of our life. Although there are many other possible culprits, these are common long-term culprits leading to our unhappiness and stress.

Whether stress blindsides us suddenly or wears us down gradually, we react physiologically. We can't do anything about that, but after that initial wave of stress has been transmitted through our body systems, we fall back on coping mechanisms. Some of these are strategies that have worked in the past, and we lean on them again to get through tough emotional times. Others are newly found ways to deal with the mental chaos in our heads. Either way, we know whether it is a good or bad way to deal with the current issue.

In chapter 10, I discussed a pathway for success in dealing with problems that may arise on either of the home fronts and how we can preempt or minimize the damage caused by unfortunate events. But now we have arrived at one of the most difficult topics, especially if it pertains to you personally: what happens when we make poor choices to deal with our pain.

An Easy Cure

The pain and suffering we internalize eventually needs to be dealt with. Everyone wishes deep down that they possessed the magic formula to find happiness again in lives they have identified as troublesome. Some of our brothers and sisters may have made the decision to turn down dark paths to deal with the emotional stress that has become too much for them to handle on their own. According to Addiction Center, 29% of firefighters—or almost one in three—use alcohol to deal with their anxiety and stress on the job.[4] Whether it is drinking, drugs, eating, gambling, or any other temporary fix, the salve we apply to these wounds doesn't make them disappear.

In Jake's scenario, he fell into a pattern of drinking that eventually started to derail his career and well-being. Each source of stress hit him from a different side:

- On the one hand, Jake had fallen into a pattern in his life where he lacked a stable friend group or social support for many years. This laid the foundation for him to scramble to find a common link with other people. Once he identified drinking as a solution to this drought, he began to integrate alcohol consumption into his life almost daily.
- On the other hand, we're not exactly given the specifics of the call for which Jake received accolades a few years ago. The child he rescued was gravely injured when Jake pulled her out of the bedroom that night. Jake had never mentioned it, but the call deeply bothered him for the next couple years, and he never dealt with these memories fully. As a result, Jake discovered that drinking not only helped his social life but also helped him to forget the specifics of a horrible night where he was regarded as a hero over a devastating grab. He was caught in a crossfire between accolades for heroic behavior and the reality of what happened to the child during her rescue. He couldn't cope with both perspectives at the same time.

Drinking slowly became a part of Jake's routine. His situation and the subsequent cautionary tale about his decline at work are far from unique. Often we let our guard down just long enough for something bad to enter our lives before we identify it as a threat to our well-being.

What was lost in the shuffle of the entire situation is the way Rob handled his role in the scenario. His technique as Jake's lieutenant was spot-on—textbook officer maneuvers. He recognized the threat Jake posed on scene and dealt with him appropriately. He did not get confrontational and instead he postponed any further interaction until the right time. When the time came to call Jake out for his potential problem, Rob was kind but direct. In their conversation, he was able to flush out Jake's addiction and reassured him that the department was on his side moving forward.

There are many reasons people make poor decisions when dealing with the stressors in their lives. This doesn't apply only to the fire service; people from all walks of life wage war against demons they can't exorcize on their own. The following are just a couple reasons:

- *Instant gratification.* Many of the ways we indulge to deal with our problems give us immediate relief, if only for a short

time. Deep down we know the long-term solution hasn't been found. But for that small amount of time, it numbs the problems temporarily.
- *Availability.* For the most part, many of the wrong coping mechanisms are readily available to us. Take alcohol, for example. We live in a time when it seems like everywhere you go, alcohol is not far away. Society has become so accustomed to its presence that it is an automatic accessory at functions, parties, events, dinners, cookouts—you name it. The same holds true for almost every other vice we may have, except illegal drugs. Sadly, they too can get their claws into people we know and care about.

Here are some other common threads that unhealthy choices have:

- *Misconceptions.* What we see and how we interpret it is critical, even as we grow older. The residual memories of prominent people in our lives when we were younger sometimes trigger us subconsciously to follow in their footsteps, emulating a lifestyle we felt was acceptable. Our judgment at the time may have been clouded by false assumptions, though. People can make impressions on us that we don't realize have a negative side until years down the road; there could be a lineage of users in our family whose footsteps we follow.
- *It avoids work.* Choosing a destructive path of coping takes a lot less thinking instead of sorting through deep-rooted feelings and their origins. The preconception of healing properly always looks more appealing than undergoing the actual healing is. So we procrastinate. Until we are ready, we will never make the decision to deal with matters in a better way.
- *We call the shots.* This is where we fall into a trap of sorts. The very issue that has the potential to damage our lives is the one thing we think we have control over. Some of us may think we have zero control over situations and we have become puppets in the order of our lives at home or work. But when we indulge in these activities, we become the puppet master. We control the intake of the substance or the tendency to sustain us.
- *A slow descent.* One common quality among addicted people is that they did not start out that way. Whether they decided to embark on the downward spiral intentionally or the behavior crossed over from diversion to controlling necessity, their behavior evolved to the point of reliance.

Remember, there is a preconceived notion that negative behaviors and their reasons are linked exclusively to tangible substances such as drugs or alcohol. This is not true. A vast majority of the dysfunction we endure is behavioral.

Always a Way out

Many of us (myself included) have been impacted by events that we have tried to bury unsuccessfully. The wreckage these events thrust upon us creates behavior patterns that we don't see. Many times, this results in very subtle personality changes that involve disproportionately large doses of sarcasm, anger, sadness, and unprocessed grief. These seep out of us in small increments, and the recipients of these are people close to us. They are left to decipher exactly what it is that has changed us, without knowing the full story. Talk about being behind the eight ball! It is an unfair hand they have been dealt, and we need to be cognizant of these symptoms when they occur.

The overall result we desire is to be as healthy and as responsible as possible. However, many times, circumstances result in our settling for something less, and we are content with the label of being "complex" because of the career we've chosen. Yet it doesn't take complexity to be a pain in the neck to our friends and family; it takes a lack of self-awareness and a lack of the wherewithal to roll up our sleeves, do the self-analysis we need, and right the ship for the people we care about. In other words, it starts with us.

Above all, remember that you are never, ever alone. Find a positive sounding board if you don't already have one in your life. There is a likelihood that people in your own department have been down a similar path and are eager to help. We all need each other at some point in our journey together—nobody is perfect.

You deserve to be free from the potential dependency. You deserve to find the balance in your life. Talk to the people who want you to be happy and healthy. There are more of us than you think. If you are down the wrong path already, it isn't too late. Seek out the level of help that you think is appropriate for your situation. You know you. You also know how much fine-tuning your life needs. When you work on yourself, even a little, you have wonderful opportunities to make strides with your life and to find the balance again.

Talk to your friends at work and confide in officers you trust. If you believe (as I do) that we are all members of a lifelong brother or sisterhood, it shouldn't be hard to find someone to talk with. There are numerous programs that offer sound pathways back to where you want to be. Fight the urge to put being proactive on the back burner. Change will gather momentum, whereas the problem becomes that much harder to defeat the longer you wait. You can do it!

The following are great resources to get on the road to recovery if you feel alcohol or another substance has started to influence your life in a detrimental way:

- *IAFF Center of Excellence for Behavioral Health Treatment and Recovery.* Call 855-900-8437.
- *Substance Abuse and Mental Health Services Administration.* Call 1-800-662-HELP (4357).

People are always there to help if you reach out—if for nothing more than to hear a friendly voice of encouragement from time to time.

Reflective Prompts

1. This may be the most difficult chapter for you to engage in self-reflection. Work will be required in order to make mental inventories for ourselves, and that can make people uncomfortable. Think back to a call that you associate with one of your senses. It doesn't have to be a particularly damaging one emotionally, but reflect on one that sticks out in your sense memory.
2. Are there benchmark calls you have logged into your memory?
3. Take a thorough inward look and ask yourself if you are holding on to anything from any of the things you've seen over the years. Did you deal with them in a healthy way? Did you deal with them at all?
4. What is your coping mechanism when it comes to particularly hard calls, and how did you suppress or get rid of painful recollections that have affected you profoundly? Was your method healthy?

Endnotes

1. Mohd Razali Salleh, "Life Event, Stress and Illness," *Malaysian Journal of Medical Sciences 15*, no. 4 (2008): 9–18.
2. Robert J. Chalmers and Jeffrey L. Alexander, "Mental Health of First Responders" *Journal of Emergency Medical Services*, December 14, 2023, https://www.jems.com/mental-health-wellness/mental-health-of-first-responders/.
3. Molly McDonough, "The Connections Between Smell, Memory, and Health," *Harvard Medicine*, Spring 2024, https://magazine.hms.harvard.edu/articles/connections-between-smell-memory-and-health.
4. "Addiction and Emergency Responders," Addiction Center, last updated June 13, 2024, https://www.addictioncenter.com/addiction/emergency-responders/.

12 All Shapes and Sizes

Teamwork. A few harmless flakes working together can unleash an avalanche of destruction.

—Justin Sewell

A Fish out of Water

What the hell am I doing here?

The thought kept racing through Kerry's mind as he was getting pushed back by a fire that was gaining the advantage on him and his colleague on the tip, Stu. Stu was a 50-something overweight country boy. They were inside a two-story farmhouse that was originally a room-and-contents fire that had ran up the wall and over their heads into the second floor.

Kerry was 25 years old but still green by department standards. With aspirations of being a firefighter in a large metropolitan department, he sat for several big-city tests and came up short on all of them. After shifting his attention to the suburbs, he landed on a department with about 20 full-time members. The move got his foot in the door with a department. Although it wasn't his initial plan, this had gotten his career going. That was 4 years ago.

After the excitement of his first full-time position wore off, Kerry began to settle into his routine. The pay at the department was midrange for the suburbs, and the fires he saw were few and far between. The department itself saw its share of structure fires, but he always seemed to be off that day or on an emergency medical services run that took him out of the action.

After his first year in the department, Kerry met Angela. They fell for each other and really connected. After less than a year of dating, he proposed and she enthusiastically said yes. Like they say, when you know, you know.

Kerry and Angela were married the year after that, and as a wedding gift, her parents gave them money for a house and a sizable piece of land. However, this posed a hard decision: Unfortunately, the land was three counties away from where Kerry worked, and Angela had landed a very good job out of college as an accountant locally. After numerous discussions, they decided to pick up and move to the country. Kerry admitted that he enjoyed his job as a firefighter but remained undecided whether it was truly his passion. But Angela was his real passion.

After they moved to the small rural town together with virtually no housing expenses, Kerry decided to throw his hat into the ring as a volunteer with the local fire department. He wasn't sure exactly what was in store for him but embraced the challenge, deciding it would be an adventure.

Kerry was hired on the spot, but the volunteer department left a few things to be desired in his eyes. There was one overnight member who slept at the tiny station. The station had exactly one engine and one emergency medical services squad. The gear he was issued was outdated and a little snug. Their air packs were functional at best. The equipment on the rig was worn and old.

The chief, Aaron, was great. He had lived in the town all his life and took extreme pride in his work. He made Kerry feel welcome and accepted. The firefighters were terrific as well. They worked hard and believed in serving the small town to the best of their abilities. While other departments Kerry tested for were asking for background checks and work experience, the small department was asking to put a makeshift siren on top of Kerry's pickup truck. This was all foreign to him.

So here Kerry found himself when the call for a house fire came in, with Stu being the only firefighter on station watch that night. Stu drove the engine, and they were the first to arrive on the scene. No cars or lights were visible, and there was farmland as far as they could see on either side. Both donned their gear and flaked out the hose for what Kerry thought would be a quick hit from the front yard. But Stu had other ideas: He quickly threw the engine in pump gear and bled the nozzle, then set it down and prepared to go on air.

Kerry froze. He had his personal protective equipment ready, but never in a million years did he expect what was about to occur: He was about to leave an unattended pumper with no support in sight and an offensive attack imminent. He eventually snapped out of it and followed Stu's lead, masking up, going on air, and entering the burning house.

The what-ifs rushed to Kerry's head—along with the anxiety. *Where is the mutual aid? What if there is a collapse? Have they never heard of waiting? What about a 360° evaluation? Was this even worth going into? There's nobody manning the pump panel! What if we get a kink in the hose? What happens if the tank empties? We have no hydrant! Is anyone even home?*

These thoughts were coming at Kerry in a rush, and he was beginning to panic. The only thing that grounded him even for an instant was Stu's confidence and determination.

"Heck yeah!" Stu yelled through his mask. "Let's beat this old girl back!"

Stu grabbed the nozzle. Kerry picked up the line behind him, and they entered the house.

Stu, sweating in the heat, glanced back at Kerry and yelled in his general direction: "Stay close to me. We'll hit her hard from the front door. I don't think it got upstairs yet."

Kerry weakly nodded, wanting to vomit in his mask. They entered through the front door with black smoke starting to bank down, but visibility was holding on. Kerry stumbled trying to keep up with Stu as they raced through the smoke and toward the seat of the fire. Visibility was beginning to get worse. Kerry remembered that under no circumstances was he to let go of the line—the lifeline back to the front door.

Stu opened up the tip when they found the living room well involved in the Charlie-Delta corner of the first floor. Here it got noticeably cooler, and visibility turned to nothing.

"Gimme some hose!" called Stu. "We gotta push this thing back again. There's more in the den up ahead!"

Kerry fumbled with the hose that had hung up around one of the corners leading to the living room. He managed to get another decent amount in the door and find his way back to Stu, who seemed possessed with purpose.

They advanced into the den to see more fire, and Stu opened up on it again. He crushed the thermal layering toward the top of the walls and then slammed his stream into the ground in a sweeping motion. Things seemed to be going their way until Stu shut down the line and pointed upward. Kerry could barely make him out through the smoke.

"Ah, she vented herself. She got upstairs through the roof!"

"Now what?" Kerry managed.

Before Stu could respond, fire had crept behind them and rekindled severely with a partial collapse. Their escape back toward the front was blocked. The hose was immobilized, and Kerry was about to freak out.

"Now what?" Kerry yelled again.

He glanced at Stu, who had opened the nozzle again and hit the fire behind them, until it began to spit and sputter. They were running out of water.

Kerry glanced in all directions, becoming claustrophobic. He laid eyes on who he was convinced would be the last person he would ever see. But Stu was already in motion. He had found a large end table and smashed it through the window on the far side of the den, like a gorilla hurling a rock. It crashed violently through the glass, and Stu motioned to Kerry to come his way. It was startling how methodical he was in his actions.

Kerry arrived at Stu in full fight-or-flight mode. Stu grabbed a leg that had splintered off the table and cleared the window quickly.

"Out you go," Stu said calmly.

Kerry dove headfirst out the window and into a garden. He got tangled in some chicken wire briefly after he rolled away, but at least he was out of the house. Stu deliberately stuck a foot and an arm out the window and then pulled the rest of his large body through the opening, landing on one knee then rising and walking briskly away from the heat. Kerry trotted to him to catch up as the two circled around the Delta side and back toward the front yard.

By that time, four pickup trucks (all with sirens), another engine, a tanker, and about 10 firefighters had arrived on scene. Aaron walked over, meeting Stu and Kerry in the front yard.

"Hot one, Stewie?"

"Lil bit. Got behind us. The kid did good, though."

Kerry had removed his helmet and hood, looking pale and sweaty.

"How 'bout you kid?" Aaron asked. "You OK?"

Kerry nodded, winded. He walked over by a clump of trees near the engine, threw up, and went to get some water from an old Gatorade container that was on the tailboard of a black pickup truck with red running lights illuminating the ground underneath it.

"Good work," a voice said behind him. It was Stu, lighting a cigarette and looking back at the house. It was slowly starting to simmer down as two other crews were blasting it with two other handlines.

"Do what you can with what ya got. We went in and did what we needed to, but she got the upper hand."

Stu put his hand on Kerry's soaked shoulder.

"Sometimes she wins the battle. Just make sure she doesn't win the war."

Kerry went over to get a chocolate chip cookies one of the families had sent over with another of the firefighters. It was the best cookie he ever ate in his life.

Different Strokes

The balance we find is a living thing. It is fluid in nature and ever flowing within and around us depending on the circumstances we run into. We've already seen how it intertwines in the fabric of our daily lives, our work lives, our home lives, and our lives in general. We must identify the balance in many of the factors we face at different points in our career and the distinguishing factors that separate each kind of department.

Kerry wasn't given much time to adapt to his new department, nor was he given any warning that the tactics used in the much smaller rural town would be a distinct departure from any of the stations he had been exposed to previously. Still, the tactics used were effective and had a balance to them for his counterpart, Stu. Prior exposure to these tactics might have equipped Kerry better for what awaited him at the fire. Instead, he was thrust suddenly into a situation that was completely foreign to him. Consequently, a sensory and emotional overload was triggered.

Kerry's story, although dramatic, represents a microcosm of what happens to us all throughout our daily lives in different departments. Whether we are part of a large metropolitan department, a suburban medium-sized firehouse, or small volunteer organization, we must adapt to the culture and recognize that we are part of something bigger than ourselves that deserves our attention to details.

The Big Three

I call the three types of fire departments the "Big Three" because almost all fire organizations fall into one of them. We must prepare ourselves for what lies ahead and put on our emotional and decision-making armor to achieve the balance in whichever one of them we find ourselves.

Large Metropolitan Departments

The largest departments of the big cities are well-oiled machines. We integrate into their system when we are hired and have to remember that most of these big-city departments have been around forever. Thus, they have a way of doing things, and as a member of these behemoths, you are now a part of that lineage, history, and pride.

You are also part of a process. Well-established tactical and cultural practices are followed by the hundreds of current—and thousands of former—members whose footsteps you are walking in. This is one of the best places to find a brotherhood immediately, but at the same time, it is the toughest of the three types of stations in which to make a huge difference on an individual level. Each station around the city has its own way, its own vibe. Some of these welcome new members with open arms; others are set in their ways and take in new cadets because they have to, rather than because they want to. This can pave the way for an easily found balance or a troublesome pathway that can rapidly spin out of control, respectively.

One easily apparent way to be rapidly recognized in these departments is to come in swinging. However, cocky, attitude-driven members will be put in their place quickly, and the outcomes for them are usually not good. Most of the time, a reputation will follow a young firefighter from house to house. Be careful when entering a large department. They are looking for someone with thick skin and a knack for learning and growing. Let the energy of the house flow to you instead of imposing your energy on the shift. Read the room and determine where you can forge friendships and gather insight. Even the most hardened and battle-tested senior members want to be respected and called upon for the knowledge they can impart.

Each department's size has its own strengths and weaknesses. Large metropolitan departments frequently have more structure fires because of population densities that exist within their jurisdiction. This is relevant because many times these departments will force you to hit the ground running with your tactical prowess and emotional fortitude. The balance will come with these departments, but you will be regimentally taught how to think, act, and perform to the best of your abilities. Those will all be mapped out for you in your academy and your specific station. What the department won't do for you is teach you how to cope. Don't get me wrong: Support resources will be in place, including excellent employee assistance programs to help brothers and sisters going through tough times. But it is an individual decision when you feel you need these programs.

A lot of times what happens in huge departments is that we put on the act of being in control when we really aren't. Down the slippery slope we slide, and before we know it, we are testy at home, abrasive in the field, and insensitive to everyone around us. Even worse, we may turn to a dark coping mechanism while suppressing these feelings (as discussed in chapter 11).

Candidly speaking, it is imperative that we monitor our progress toward the balance in the large departments. A sea of members, a sea of fires, a sea of death and tragedy, and a sea of emotions exist in the vast majority of them. Stay focused on yourself and be sure that you are practicing the critical skill of transparency. Do this not only with your coworkers but also with officers, your family, and above all, yourself.

The following are suggestions to help find the balance:

- *Be teachable.* And not just as a cadet fresh out of the academy. At any point in your career, you can learn. Processes and technologies in the service are always changing. Be open-minded. Even the most stubborn departments in the world will adapt to save lives, time, and money.
- *Be approachable.* We aren't making widgets in a factory or doing mundane tasks in an office. We deal with life-or-death situations.

If someone hesitates to ask you a question they think you may know the answer to, it could cost them—or you—dearly. Let them know what you offer right out of the gate. It works better and you don't sacrifice an ounce of reputation to do it.
- *Keep within your bubble.* Imagine you existed in an impervious bubble where any emotional jabs or abuse had no effect. Would you be more relaxed? Of course you would! So, use one. There will be members who challenge your readiness with verbal battles and cutting remarks. Be ready and know they will inevitably arrive. The key is to not let them determine your self-worth afterward. If a member or ranking officer belittles you, remember it is only in your own head. Don't give it any energy. You are the battery pack that supplies energy to whatever emotion you feel each day.
- *Don't take anything personal.* We are doing a job and running at a high-octane pace in the large cities. Toughen up. The reality of our society entails empowering employees and making sure everyone is respected equally. Nevertheless, being a firefighter in a big city is a challenging combination of changing with the times and steadfastly keeping in line with department traditions. Again, get tough. Just don't turn bitter over it.
- *Trust the process.* You may have reservations about your decision to join a large department. Making this leap may or may not be for you. Trust your gut and figure out whether you are simply going through growing pains or trying to fit a square peg in a round hole in the long term. That self-reflection will serve you well and save you years of frustration if it turns out you aren't cut out for a department of this size.

Suburban Departments

The second type are the suburban or midlevel departments. A vast number of these are spread across the United States, and they usually consist of a central station with or without smaller satellite stations.

In these departments, you can encounter a large array of organizational stages, work-in-progress hierarchies, and daily occurrences that are broader in scope than those in the other department types. Why? Because suburban stations are not as fragmented and spread out as metropolitan departments while also tending to be more cohesive and routine driven than the smaller rural departments.

Suburban departments are their own entity. When traveling across a large metropolitan area, you may stop in and find subtle differences between stations just 10 blocks away from each other. Yet the same framework exists, and they all answer to the same chief, mayor, and council. In contrast, two adjoining cities with suburban departments could have a completely different way of doing things: different protocols, different cultures, different everything sometimes.

As a firefighter or officer of a suburb, you will find everything concentrated in the central station, which acts as a nerve center for the entire operation. Most often, this is the one type of department where the chief or chiefs sit literally feet away from the shift members. This can try the patience of everyone involved. Hence, for the balance of the entire department, it is essential that each member stays in their lane for this setup to work efficiently. Delegation, accountability, empowerment, and not micromanaging will be the only true pathway to success and the balance here.

Because suburban departments tend to be more of a pressure cooker in relation to city hall and fire service administration, the shift members can easily feel as though they are under the microscope while the administration's decisions are scrutinized and categorically judged. This comes with the territory. Close quarters mean a front-row seat to decision-making.

There is still a job to do, rules to live by, and a balance to be found. Although there are slight differences, the basic framework for success remains the same: You're still a firefighter and still expected to perform your job in a professional and efficient manner.

Recommendations for suburban departments include the following:

- *Embrace the tight-knit group.* You are part of a family at the firehouse. Like all departments, you have coworkers you refer to as brothers and sisters for good reason: There are instances where you put your life in their hands, and vice versa. You must realize that inflaming petty squabbles around the station will fester in a fishbowl—in other words, there's simply nowhere else for the tension to go. Make the most of what you have and let the little things go. You have no idea what the trickle-down effect of tension between members will be for everyone else.
- *Training has an endgame.* Because suburban departments typically have less actual structure fires than their metropolitan brethren, training becomes incredibly important. Senior members may scoff at the notion of training all the time, but it matters. Repetition breeds habit. After all, would you rather train on a Mayday call in a makeshift scenario behind the station or be clueless and sketchy during an actual one? Me too.

- *Trust your people.* This goes for officers and chiefs more than anyone else, but it pertains to everyone on the department: Don't just declare that you trust your people; really trust them. They are respected members in every other relationship they have in their lives, yet at times, they may feel like they are being spoon-fed tasks like toddlers. Avoid that at all costs. Don't make their reward for rigorous training and stressful testing be that they are treated as incompetent children who need constant monitoring. It's insulting to them and exhausting for you.
- *Take an interest in others.* Anyone can show up for work, punch a clock, do their job, and leave. Put thought and energy into forging relationships with those around you. Something as simple as saying "good morning" to a coworker goes further than you think. Even when you have 20 other pressing concerns on your mind, being cordial with others works wonders.
- *Accountability is paramount.* A suburban department doesn't have any more or fewer criticisms than other departments. The problem, though, is that because most of the members are in such close proximity to each other, word spreads like wildfire, and almost *everything* is being mentally recorded in one way or another. Embellishments of stories are horrible. A harmless remark can be twisted into a harsh criticism after filtering through several links in the chain from source to listener. Hold yourself and your peers accountable for the comments they make. In turn, search remarks for clarity to keep a minor misinterpretation from turning into a full-blown misunderstanding.

Rural and Volunteer Departments

Rural and volunteer departments comprise the last type of departments we may be a part of in our career. Small departments like these have a unique dynamic: The members of these departments are typically close in values and work ethic. They enjoy each other and work well together. The largest difference with these departments is that they rarely share a living space and only convene during emergency situations or on occasion for training. On the one hand, the bond formed by living together doesn't exist; on the other hand, members typically don't have exposure to each other long enough to get on someone's nerves.

Returning to our story at the beginning of the chapter, Kerry was exposed to a clash of tactics. He had learned basic skills most of us acquire in academies and tried to apply them as he limped through sporadic fires on the suburban department he was first hired on. But after he joined the volunteer department, things changed.

Volunteer departments sometimes rely on a critical trait not many others do: ingenuity. Because of the lack of adequate personnel and equipment, firefighters must make do with what is available. In the scenario, Stu showed experience in fire behavior and didn't think twice about the catastrophic possibilities of leaving a pumper on its own while going into a fire. Kerry was along for the ride and was blown away. As the scene played out, Stu advanced the line and chased the fire with what seemed like little regard for the danger they were facing.

In the end, Stu was able to recognize the drastic conditions confronting them and decided to exit before matters got any worse. A less experienced firefighter might not have even attempted an interior attack. He drew on instinct and went forward. Kerry was taught an important and near-fatal lesson in his eyes: Be prepared for the unexpected when crossing over to a different department.

Typically, when we head someplace new, there will be an established training regimen. In Kerry's example, he wasn't really afforded an opportunity to acclimate himself to the new department and their tendencies. As the scenario demonstrated, decisions were made that ran counterintuitive to what formal training dictates. Sometimes this is the case on smaller departments. The men and women at these departments can be exceptional. Unfortunately, they can rapidly become cautionary tales of tragedy and statistics as well. Necessity breeds invention. At times, it also breeds bravery, sacrifice, and poor decision-making.

Most volunteer departments afford you the luxury of staying connected at home. Because you are not gone for a full 24-hour shift, you must be ready to flip the switch and get your mind ready to go in a flash. It is critical that you find the balance in a volunteer department. Whatever works for you will pay dividends if it gets you in the headspace you need to go from a backyard barbecue to a full-blown ripper in a matter of minutes.

Protocols and procedures are usually passed down through an informal relay of messages if meetings and training are not held regularly. Many times, tactics lean less on streamlined policies and more on individual preferences. Make sure you are on the same page to keep consistency during emergency situations.

The following suggestions may prove beneficial as you navigate membership on a rural or volunteer department:

- *Communicate clearly.* Communication is huge in the rural arena. We've touched on the necessity to communicate across both sides of the fence between your two families. This, however, pertains more to tactical interventions. Granted, the story at the beginning of the chapter seemed rather extreme in the offensive operations implemented, but sooner or later, the tricks of the trade will come out of a pocket that may catch a fellow member off guard. Although communication isn't a catchall for everything that could occur, even a 10-second debriefing prior to entering a hazardous situation can pay off in large dividends.

- *A training program is essential.* You can check many boxes on a volunteer department if you have cohesive training. As a chief or senior officer, this can show you exactly what you have to work with and can rally the troops. As a shift member, it will give you confidence emotionally and settle you down in crisis mode. Poor Kerry was subjected to a perfect storm. He had no idea how a rural department worked in order to succeed. Risks typically considered not prudent were taken for the sake of solving the problem at hand.
- *Be independently dependent.* One of the great calming factors if you are working on a small department with limited resources is to take it upon yourself to hone your skills. When you have decided to give a small department your loyalty, give it fully. Make yourself the best you can be and strive for excellence. Demonstrate that excellence in everything you do. Chances are it will rub off on those around you, and everyone will get better.
- *Know your peers.* It can be difficult at times to find the balance with others you work with and know at the station. It is impossible to find that balance when you know nothing about your coworkers. If you see them only during emergency runs and a yearly spaghetti dinner fundraiser, the balance will be unattainable because you have no idea who you're working with. Get to know your coworkers. Discover their tendencies. Play scenarios out in your head and ask them what-if questions. You'll be amazed at the amount of information you can deduce by asking simple questions.
- *Stay fresh.* Volunteer departments are the polar opposite of large, big-city ones. You have an opportunity to make your mark much quicker in smaller departments in an impactful and positive way. Chiefs are always eager to make their departments stronger, and they usually have an extreme sense of pride at rural departments. Keep a fresh set of eyes on everything around you and realize that cost-effective improvements are always possible. The balance will come when you are aware of your place in the order

of things, and you will earn respect from others and yourself when you avoid complacency and always try to improve.

It's Your Time!

When we hear this battle cry, we may think of everyone rising in unison and seizing the moment to define ourselves during a crisis. That isn't exactly what I mean, though; what I mean is that we are all given a finite amount of time, so don't waste it. What you choose to do with this time is up to you. Try to find the balance points in everything you do, regardless of the size of the department you are on. When you get mentally lazy, you lose focus and become stagnant. Take breaks if you have to but always move onto the next thing you want to get better at. You'll find your stay at the department you've chosen will be much more enjoyable that way.

Reflective Prompts

1. Most likely, you are about to be hired on a department or are currently on a department. Regardless of whether you have found your groove already, there is always time to take stock: How is it going for you? Do you feel you are behind, ahead, or right where you envisioned yourself being given the department you are on right now?
2. If you have yet to get hired, what goals do you have in the first 3 months? The first 6 months? How about a year? This can apply to any phase of your career you are in. Whether you are a senior member or just starting out, always have goals to better yourself.
3. Do the department types detailed in this chapter coincide with what you have experienced? How would you tweak them to fit your situation better?

13 That Darn Ego

The best way to live a miserable life is to pay attention to what other people are saying about you.

—Paulo Coelho

The Eyes of the Devil

The dirty plates were still on the table after dinner on a Saturday evening. The younger firefighters were hanging on every word from Smitty and Bud. They were recounting in great detail "The Story of the Walker Road Save," a legendary fireground performance they had been a part of one rainy spring night 12 years ago.

The department had 50 full-time members and was growing. Elton Smith, known affectionately around the department as "Smitty," had just completed his 23rd year on the department and was on shift, making a rare weekend appearance.

Smitty's partner in crime for over 15 years was none other than Gary "Bud" Turgerson, a reckless outlaw to everyone on the department. He had a silky mustache and an abrasive demeanor. That night, Bud was working the shift as a favor for a midlevel firefighter, in repayment for a trade over the holidays. He was coming up on his 21st year of service. Both men were legendary for their acts of bravery, and younger members could only wish that someday they would have stories even approaching their adventures.

Smitty had his feet up on the table, while Bud sat back in his chair with his hands clasped behind his head. Eagerly listening were a young crew of newer members, all imagining the harrowing ordeal the two veterans endured years before.

The shift captain, Anthony, leaned against the wall, sipping his after-dinner coffee from his favorite mug. He nibbled on a cookie from a tray contributed by the parents of a Girl Scout troop that had toured the station earlier in the day. He didn't say a word but listened closely along with the rest of the audience.

"So I get to the top of the stairs—and wham!" Smitty said dramatically. "The entire roof on the second floor starts to cave in."

Bud chuckled and added, "I look at Smitty, he looks at me, and I'm like, 'This ain't no good!'"

Laughter from the captivated audience.

"So you guys were search and rescue on the second floor?" asked one of the new members.

"Yeah, aren't you listening?" Slightly irritated, Smitty continued: "The hoseline got leaks coming from the coupling, and our pressure is nothing."

"But we knew the toddler was at the end of the hall in the bedroom, and we're not gonna let her go," Bud said.

"No. So we decided to sound the floor and open up with whatever water we had," Smitty said as he took his feet off the table and leaned forward.

Bud still held the pose as he scanned the room of wide-eyed listeners. He then continued, "So Smitty finally says to radio for back up or something. I can't remember and—"

"But my radio was dead—as a doornail!" Smitty chimed in.

"And I didn't even have one," Bud added.

"So here we are, no radio, nothing for water, no backup, no idea what's going on downstairs, and a trapped baby down the hallway yelling and screaming her head off."

"Where was your radio, Bud?" asked another youngster.

Downing a swig of water and wiping his mouth, Bud barked at the kid, "I gave it to one of the new guys that forgot his."

He looked meaningfully at Smitty, thinking for a second. "I can't remember who it was, but he sucked."

"Anyways, it's getting hot and things are going to hell in a hurry," Smitty went on. "Suddenly the ceiling leading back downstairs collapses, and now it's really go time."

Bud sat forward in his chair. "It was decision time. So Smitty told me we gotta hit it hard with the rest of the water and make a run for it toward the screaming girl."

Smitty sopped one more bite of spaghetti sauce from his plate with his bread. After popping this in his mouth, he continued, "So we opened her up with the trickle that was left, looked at each other, and made a run for it."

The room was tense with anticipation of what was going to happen next. The younger members felt as though they were in the presence of true greatness and were proud to be their audience.

Bud said, "We both barely got to the door when the rest of the ceiling collapsed into the hallway. We slammed the door and started scanning for the kid."

Smitty got up from the table, really into the story now. "It wasn't hard, she was crawling on the floor next to her bed."

Bud took the narrative thread again, "So I throw the window open and start tossing stuff out the window until one of the guys outside sees us. By this time, the smoke wasn't terrible because the door's shut, but it's getting toasty up there."

Bud got up and leaned on the table. He was full-blown into the tale as well. "I've got the kid in my arms, and they throw a ladder to the window and we hand her out. They told us that the entire roof was about to go and to get out."

Bud stood up and closed his eyes as he made a dramatic pause. He slowly shook his head and opened his eyes again, as if he was right back in the bedroom recollecting every moment from that night.

"I climbed out the window and saw this guy go bailing out right past me after I cleared the opening," he said, nodding to Smitty.

"Roof caved in 30 seconds later," concluded Smitty.

Smitty shook his head and grabbed his plate. He turned back toward the table full of firefighters and gave a modest shrug. The entire room exhaled as the story ended with two heroes making a save for the ages.

The crew got up from the table to begin their post dinner chores. A few tagged along with the storytellers as they walked into the kitchen and put dishes in the sink.

"That story is insane!" one of them said.

"Yeah, no thanks. No water, collapse—no way!" another one chimed in.

"Two in, two out," Smitty said. "Gotta know whoever's with you has got your back."

Anthony didn't say a word, putting his mug in the sink on his way up to the shift office.

Later, Bud walked by the shift office and poked his head in the door. "What'd ya think?" he asked Anthony.

"Bravo!" Anthony said, still looking at his paperwork. He took off his reading glasses and looked at Bud with a smile. "You actually had me going there for a minute."

Bud smirked and shrugged, guilty as charged.

"That story gets better every time I hear it."

"We may have talked it up a tad."

"Well, leaving out that the kid was 12 and climbed down the ladder on her own and the second floor never actually collapsed was a nice touch."

"It collapsed a little," Bud said with a grin. "But the kids back there ate it up."

"Sure did," Anthony said to himself as Bud disappeared down the hallway as he continued with his run reviews.

The Rolling Eyes of the Watercooler

The embellishment of stories is legendary in the fire service. Who doesn't like to pepper marginally mundane calls with interesting dialogue and exaggerated actions? It's *awesome*. I equate our tales to those that hunters come up with around the campfire. What starts out as a button buck eating in a field turns into a trophy elk charging at us 10 years later. We love hearing and telling good stories.

"The Story of the Walker Road Save," like many of the fictitious tales throughout this book, was exaggerated by Smitty and Bud for dramatic effect and to prove valuable points. The pair wove their yarn in unison, hooking the listeners and building up to a resounding climax. The tension of the young members in the audience was palpable. But like every good story, there was another buildup that took place.

In our field, whether we like it or not, our reputation precedes us. The danger, the intrigue, the reliability, and even the calendars portray us as real-life heroes on a daily basis. And in many ways we are: It's OK to take pride in doing important work, and we certainly deserve to. But the ego—which we cannot see, cannot feel, and cannot detect until it is threatened—can be one of the ugliest things a human being possesses. Once it is awakened, it will protect itself at all costs if not kept in check.

The game the ego plays with us is astounding. Even when we think it is not in control, it lays in wait. People in the private sector have it as well, usually triggered in the contexts of social status, wealth (a big one), family pecking order, and reputation, among others. In general, our egos are linked externally with our job, internally with our reputation.

In the fire service, we also must contend with the normal triggers that afflict the general population, as well as vanity. Our egos give us a feeling of control but only in short bursts; moreover, the ego is responsible for a great deal of grief in our lives. During emotionally overwhelming events, such as funerals, births, and life-or-death situations, we cannot control our ego. In these instances, our deeper nature is exposed because we can't feed the ego any longer.

On a less threatening level, we see examples like our story. Smitty and Bud were inflating their egos by engaging the younger members with the story they were telling. The younger members couldn't get enough, and neither could the storytellers. Their egos were being fed because they knew there was no way either of them could be looked at as anything other than tough-as-nails firefighter heroes. The ego is the little part of us that acts as a gatekeeper, constantly monitoring what will make us look better or worse in the eyes of others and ourselves.

Thus, we build ourselves up and guard against threats to our egos (which are tied very closely to the roles we play). We continue doing this until something beyond our control comes along and challenges our ego's energy. Even so, the stories will always be fun to tell.

The Two Victims of the Ego

There are two main victims when the ego rears its ugly head and takes control of us. It can have either an impact on others or an impact on us. Both have powerful consequences, and these can unfortunately overlap.

The Ego Against Others

The brutality of the mind's ego can negatively impact everyone around you at work or at home. When we decide to move a certain way with our relationships and people's perception of us, the first stone is cast. We are no longer an open book, and we self-label ourselves one way or another.

There are several different ways we may do this. First impressions rule highly among these: Once people have catalogued us in a way we think they do, we then play that part in an ego-driven game. A great example of this in society is posting on social media. People will expend great energy to achieve the perception of how well they are doing and how great their life is. This is a powerful way to feed the ego in quick fashion. We touch hundreds or thousands of people simultaneously with great news of our fascinating lives. Why? To make them happy? No, it is to inflate our own egos and validate it in front of the world.

In the fire service, we have invested a large amount of time to build a reputation, and our ego is always present at work, regardless of whether we recognize it. We tell stories like Bud and Smitty do, put others down, take advantage of another person's weaknesses, and avoid embarrassment so we don't seem vulnerable to other members. We've all done it. We just don't recognize it. A good example at work would be to avoid a piece of equipment we should be proficient on in an effort to not look foolish in front of other members or officers. Nobody likes to look dumb, especially not the ego!

Another classic example of the ego at the firehouse is an interactive one. An almost universal affliction exists to one degree or another with regards to verbal jabs in our industry. In chapter 3, we met Bryan and Cody, two midlevel firefighters who chose to give the new hire, Jeremy, a hard time. It wasn't until Ray, the senior member in the scenario, came in and disarmed the tense situation that occurred. This is a perfect example of how egos can take over the actions and reactions during conversations. That chapter, by addressing the roles that we play in our lives, overlaps with this one, which discusses how we inhabit those roles. To establish our roles, we must cater to something. In Bryan and Cody's case, their egos ruled supreme. They fiercely displayed their dominance to protect their egos.

In my opinion, the ego has one source that feeds it: fear. When we lash out at others and make cutting remarks to each other, fear is usually involved. The fear of being vulnerable in one fashion or another drives our actions. People will go to great lengths to protect the image they think others have of them. Most of the time, their suspicions are wrong, but human nature dictates otherwise.

As difficult as it is to keep our egos in check at the station, it is equally as difficult at home. Plenty of times we head home with chips on our shoulders and duct tape over our mouths. We creep through our home life by being "strong and silent"; however, this gets stale in a hurry. The people we love quickly tire of that act, and the tension eventually explodes into an argument or fight. Then the waters recede, the egos involved simmer down, and the vicious cycle begins again.

That is why the balance is so important. When we keep the communication pipeline open with transparency and respect, we can at least identify when the ego is about to come out to play. Finding the balance won't necessarily get rid of the ego, but it can certainly curb its potential nastiness. Healthy implementation of good habits will create allies on the home front to battle together, instead of adversaries left guessing why we are acting the way we are.

The Ego Against Ourselves

Public speaking is the number one fear of people in the United States.[1] Death is number two. Let that sink in. Why do people have such anxiety over speaking in front of others? It's because of—you guessed it—the ego. We are so afraid of looking a certain way in front of others that we will do anything to avoid it. But did you notice what perpetuates the anxiety? Fear. Social situations may cause so much fear that some people get violently ill or faint. This is because they are wound so tightly around their ego's little finger without even realizing it.

We play mind games with our egos every single day in a wide variety of situations. We defend our egos at the gym, at the grocery store, in traffic, in front of our kids, at the beach, on vacation, at family gatherings, at the firehouse, while pumping gas, with our spouse, even in chat rooms online! People can't even see us, yet we still must cater to the ego's hunger to be right and look superior to them as they read our posts. It is insane—a disease. The ego dictates our actions in a never-ending fashion.

A good way to test the ego's hold of you is to consider how you feel wearing different clothes throughout the week. Do you feel the same when you have a classy outfit on for a wedding versus your bunker gear headed into a fire? Imagine someone watching you objectively who doesn't know you personally. Would you have a different swagger when entering a room in your Class A uniform instead of, say, a T-shirt and jeans? Most likely, the answer would be yes. Say hi to your ego, because that is why you feel that way.

There is only one true defense to the ego in your head: being aware. When you are aware of the monster in the room, it loses its ability to reach up and bite you. Keep an eye on your ego. It can be lethal to your well-being if you don't.

We work in an industry where our reputation precedes us. It's not uncommon to walk a little taller and hold our chin up a little higher when people know what we do when we arrive on scene to save the day. Don't get me wrong, having pride in our work is vital for emotional fulfillment. But being a self-absorbed, pompous, condescending jerk is not. Other occupations share similar challenges including people whose occupations yield great wealth, such as doctors, public figures, celebrities, and sports figures. Let's pump the brakes and realize we are all just human beings doing human being things. Be aware. It matters.

Breaking the Cycle

How can we break the vicious cycle comprising our egos and their hold on us? Here are a few helpful tips:

- *Watch yourself.* This is extremely important. Take note of your behavior from time to time and decide if you truly like what you see in the mirror. You should *always* strive to be the best version of you. There may be instances where you have overreacted

to someone or said something you wish you could take back. Welcome to being human. Just always try to improve yourself.
- *Put yourself in someone else's shoes.* I'm not just reiterating the "Do unto others" motto but asking you to look at how you are affecting others. Is your ego involved in even the smallest decisions you make? If so, what are you trying to defend so vigorously? Look at questionable interactions from the other person's perspective. You'll be surprised by what you uncover, and you might realize you are exhausting yourself by putting up a front.
- *Determine whether your goal is beneficial to the common good.* At the station, is the perception others have of you a good one? Hopefully this is answered by a resounding yes. However, if you don't know—or even worse, don't care—then most likely your ego is dictating your actions. For most of us, our egos are so ingrained within our personalities that we don't even recognize its presence. It can take people years to separate themselves from their egos. Make sure you and your ego are in check enough to be a good resource for the shift and the department as a whole. Be humble and honest with yourself.
- *Be kind to yourself.* Let's be honest: We can behave brutally toward ourselves. Find and practice using an outlet to stay centered and maintain your inner balance. For many people, hobbies help them to discover this personal balance, and it is hugely beneficial to them to have these outlets when crisis arrives. When you pursue inner peace through whatever gives you enjoyment, you are able to center much better during difficult times. If we turn a blind eye to our body's needs when the seas are calm, we tend to fall apart when the waters get choppy. Be proactive and have something similar in place to help center yourself. As firefighters, we are *always* prepared for everything. Yet we don't stop long enough to care for ourselves.
- *Ask yourself: What are you afraid of?* Great question. It applies to almost every ego question in almost any situation. Remember, ego is fear based; thus, any imbalance in a situation is because you are afraid of something. One great exercise, which doesn't have to be in the middle of a stressful encounter with someone else, is to try being self-deprecating when telling a story to the shift. Can you do it? Can you make yourself look bad to the shift, telling a story where you truly looked like a fool? This is a litmus

test to see just how much of a stranglehold your ego has on you. If you can't tell a story like that to others, learn to do so.

Three's Company

In our departments and our people, the ego will always have a place. In any industry or managerial hierarchy, the ego exists. We have to deal with it and recognize when a fun story that pumps us up in front of peers crosses over the line into biases, attacks, and all-out assaults on others to preserve our egotistic way of life. Through the suggestions outlined in this chapter, we can become aware of the ego's presence and keep it in check. Without a recognition of its existence, the balance will be incredibly hard to find.

Reflective Prompts

1. The ego is powerful in all of us. Be sure to recognize it. This chapter ties back to the discussion of roles in chapter 3. With that in mind, can you see where the ego might fuel the fire of the different roles you play in your life?
2. Sometimes the roles we play are necessary for the tasks we have undertaken as firefighters, spouses, mothers, fathers, daughters, sons, friends, and many other roles. Are you happy and content with the role(s) you have established in your life? Is the role fear driven, or does another motivation perpetuate that role? Be honest with yourself, or you will fail to be honest with anyone else. Self-assess and see where you sit. Work on the egotistic roles you have taken on.

Note

1. Pat LaDouceur, "What We Fear More than Death," MentalHealth.com, last updated September 25, 2024, https://www.mentalhealth.com/library/what-we-fear-more-than-death.

14 Empowering and Its Benefits

The best executive is the one who has sense enough to pick good men to do what he wants done, and self-restraint enough to keep from meddling with them while they do it.

—Theodore Roosevelt

Out of Order

According to all accounts of the people that knew her, Sue was a doer. She was active in her church and sang in the choir when she was able to attend. Many of the women in the congregation loved that she would organize quarterly bake sales and spearhead the coat drive each Christmas.

Her husband, Cal, owned a successful software company and usually worked from home on his own schedule. They went to the same high school when they were teenagers but never really connected until they happened to be in the same hiking club that tackled two of the longest trails in the state the year after graduation. That was a little over 13 years ago.

Sue and Cal had been married for 11 years and had three girls together: Allison was 9, Dakota was 7, and Annie was almost 6 years old. The family had a routine, as most do, with the girls already learning the value of hard work through performing chores around the house.

The couple were always active and spent much of their summers with the girls on kayaking and camping trips. Television and computer time was limited, and although the oldest was beginning to push back about getting a smartphone, they had largely been able to instill their belief in a family structure built on quality time.

Sue had two brothers and a sister. They all split time helping her mom and dad out when they could. Her parents were by no means old, but a car accident

3 years earlier left her mom with chronic back pain and her dad with a traumatic brain injury that hindered his capacity to perform his daily activities around the house. Sue and her older sister Michelle did the bulk of the heavy lifting in that department.

The 32-year-old had hobbies, friends, a family, and sound social groups she took pride in. But outside of her family, she had one true passion: firefighting. Sue had become a full-time member of a suburban department 6 years ago and balanced it beautifully into her life. Her assertive disposition and knack for keeping busy made the department a perfect fit for her. She was smart and stayed levelheaded in stressful situations. She had just one major problem in her life: Sue was not happy where she worked.

The job itself wasn't what bothered her. On the contrary, she loved it. But she couldn't put her finger on what it was that made her want to pull her hair out every time she punched in for her shift. She had excellent medic skills and was terrific on fires, as her peers attested. She tried her best to figure out why she felt so unsettled but was at a loss. The only relevant change in recent memory was the new lieutenant on her shift, Cole. He was well liked throughout the department and really sharp. *It couldn't be him, could it? He's awesome.*

Sue finally got to the point where she sat down with her old captain, Eli. They worked together for the first few years after Sue got hired, and she still confided in him even after he got promoted to assistant chief. She visited him in his office for a chat after punching out one day.

Sue knocked on the door, and Eli looked up. With a smile, he motioned for her to come into his office.

"Susie Q!" Eli said as he got up to shake her hand.

Sue strode into the office and shook his hand, giving a smile in return. After she plopped down in a chair, Eli walked over and closed his door. He sat back down and stacked some papers by his computer.

"What's up?" Eli said.

"Thanks for meeting for a couple minutes with me, Chief."

"It's no problem at all."

"So, I wanted to ask a question," Sue said, "about how you think I've been doing overall."

"You know the answer to that," Eli replied. "Everyone loves you. Never any complaints."

"What if I have one?"

"Do you?"

"About myself, really," Sue said.

Eli leaned back in his chair, puzzled. He cocked his head, trying to understand her meaning.

"Not sure what that means but go on."

Sue fixed her shirt and fussed with her collar. Eli could tell she was frustrated.

"I don't know what it is, but for a while, I just seem off."

"Keep going," Eli said.

"I don't know, but it's driving me crazy. I just feel like I'm holding myself back in some way, and I feel like I'm too young to feel so . . . unmotivated."

"Well, that's a word I never thought I'd hear coming out of *your* mouth," Eli replied.

"Right? I'm just kind of in a rut, and I can't figure out why."

"How are things at home?"

"Never better. Dakota is gonna be 8 in a couple weeks, and Annie is starting gymnastics."

"Any trouble with the guys?"

"No, we're getting along great."

"Officers?"

"You know how Nicky runs the shift, and Cole has been cool to me. He looks out for me all the time."

"Huh. There's gotta be something making you restless. Any bad calls lately?"

"Nothing crazy. No real bad ones."

Eli could tell Sue wasn't herself. She seemed irritated in a way he seldom ever saw. She was usually cool and collected. Today she seemed flat and lifeless compared to her usual demeanor. He got up and glanced out the window.

"I guess it hasn't been the same since you left and everyone got promoted."

Eli took a sip of coffee and turned, looking at Sue. The wheels were turning in his head. He walked back over to the desk and set his mug down. There was something in the timeline she just described that was making him think.

"So I don't know what's going on," Sue concluded.

"Explain what you meant when you said 'Cole looks out for you all the time.'"

"He just makes sure we are on the right track with what we're doing," Sue said, fidgeting in her chair. "The younger guys have really appreciated it."

"Do you?" Eli asked.

Sue shrugged and broke eye contact, cleaning a speck of dirt off her pant leg.

Eli filled the silence with an offer: "Tell you what. Let me talk with a few people and touch base with your officers. I have an idea about what might be going on, but I don't want to say anything just yet."

Sue fixed her gaze on Eli and raised an eyebrow. He gave that same devilish smile he used when coming up with intense training ideas when he was her officer.

"Uh-oh. I know that look," she said.

Eli smiled and shrugged. "Sit tight," he told her.

Sue nodded and walked out the door, taking comfort that Eli had a plan and a hunch.

After a week of reflecting that included a couple of meetings with Nick and Cole, Eli finally pinpointed what had happened with Sue. With Nick, there hadn't really been a departure from anything he had done before he was promoted. However, Cole's situation was a little different.

In his meeting with Cole, Eli recapped a couple of typical days on the shift. He asked Cole to walk him through what had happened each day as far as training, calls, dinner, activities, continuing education, and anything else he could think of.

Cole was precise with his descriptions, and on the surface it looked as though he was a thorough and engaged supervisor who took pride in his people and in the shift's work. However, Eli noted that Cole did an extraordinary amount of checking up on his people and did a lot of looking over shoulders to confirm tasks were being done right.

At one point, Cole sent two younger members out to do dinner shopping and called to check on them to see how the shopping was going. He had been a lieutenant for 2 years, and his style translated to one that had teachable moments in practically every task he handed down—in his eyes. While this methodology worked with those members who needed guidance, it spilled over to the ones who didn't—like Sue.

This management tactic was counterintuitive to every instinct Sue was used to in her life. She was a leader in her church and a strong family person, which included being the principal caretaker of her parents. Yet when she got to the firehouse, she was treated as a mindless drone, being micromanaged while she performed her tasks. This is not to say that Cole's style was totally wrong. But it needed tweaking.

After Eli's epiphany regarding the situation, he sat down again with the captain and his lieutenant. In the weeks that followed, Cole loosened his grip on the micromanaging of the shift. He was never berated, and in the process of the entire reformation, he was given a vote of confidence by Nick and Eli. They both met with him individually and praised his abilities. In addition, they both asked him individually if there was anything they could work on in his eyes. That opened the door for constructive criticism and an opportunity to reassure him that his management style was not essential to his success moving forward.

After his review and with minor tinkering (at Nick's blessing), Eli was able to right the ship. Sue was called back into Eli's office, and with good, strong, open communication, they were able to work out a very positive outcome that avoided hurt feelings and bruised egos.

Sometimes the balance is masquerading as something else, which we have to root out. This is exactly what Eli did. Sue was empowered to work once again independently and felt much better about herself and her future within the department as a result.

The Power of Empowering

There is no singularity in the fire service. Rather, success is gained by entities that work together to achieve tasks and to obtain the desired outcomes. When we empower the people we work with, it brings a calm liberation to the players involved. We feel as though we are part of a team instead of being told we are. Empowering each other instills confidence in one another. At the end of the day, what else is as essential when we are trusting each other with literally death on the line?

Sue's plight is a common one to varying degrees. Although some officers may not be as adamant about keeping constant tabs on their members, every department can name a situation where someone felt as though their contribution was cheapened in some way.

Although sweeping policies must be delivered from time to time, blanket decisions aren't always the best route to success. The worst thing an officer can do is to treat a room full of wolves like a bunch of sheep. We must be proactive and delve into each member's progression; we owe that respect to them, and we owe that respect to the rank. Winning a test is the easy work. The hard work starts the first day you sit in the shift office.

Developing varying degrees of empowering people may be met with hesitation by some. But give them a chance. If there is something that needs to be cleaned or reorganized around the station, send a team of people to do it and see what they come up with. Three things will happen as a result:

- First, a team mentality will kick in automatically. This is because now they have all become problem-solvers instead of order-followers. If you were a fly on the wall during an assigned task that you give your people latitude with, you would see an amazing transformation, which could possibly give you brand-new insight into their capabilities. Importantly, the members will begin to assign tasks to each other and will share a common vision that is *theirs*, not *yours*.
- Second, regardless of whether the task is done right or wrong in your eyes, it will get done. Realize that things won't always be done perfectly. That's OK. If cleaning in a way you don't really like provides data on qualities that can serve your members down the road in tougher situations, it is worth indulging their different style.

- Third, when you assign tasks without direct supervision, something will occur where you can't even see it: between your members' ears. Giving members a chance to tap into their creativity through individualized task management yields a calming effect and a feeling not only of appreciation but of accomplishment. These are huge attributes in the cultivation of strong and happy members of the department. There is a different atmosphere and a different energy that starts to transform the morale of the members when they feel like they have the freedom and power to make decisions that impact the organization they serve.

So, in what ways can we empower the people we work with? There are ways we can empower both subsets of our people in the department—namely, those you manage and those you work with.

Empowering Those Who You Manage

Let's start with how we can empower the people we have the privilege of managing. When people use the word *empower* in the workplace, these are the subset of people they are usually referring to. The following are a few suggestions tailored to this subset of coworkers:

- *Put in the work.* It is your duty to take the time to get to know each of your members. If you become lazy, it will eventually show. You will be able to hang in there for a while, but cracks in the foundation of your leadership will begin to wear down your shift. To effectively empower your members, you have to know at what level they are all operating. There's a time and a place for orders and directives, such as on the fireground. From a more personal perspective, try to find out which level each individual member operates on. Doing so will allow you to empower each one of them without overwhelming them with tasks that may be over their heads.
- *Clarify the objectives.* When you empower people in the organization, be clear on your objectives from both a narrow and a broad perspective. On the narrow side, make sure they know what is expected of them, and then give them the latitude to figure things out on their own to get to the end result. On the broad end, let them know in an efficient way that success will

breed opportunity. As their scope of responsibility grows, so will your confidence in their ability to take on larger tasks.
- *Let calculated mistakes slide.* Allowing empowered employees to make mistakes and to learn from them on their own is an effective tool. Of course, we are not going to use this trial-and-error approach to dictate results in emergent situations. But we can allow our team members to, well, figure out routine matters. This gives them an enormous sense of accomplishment that not only will contribute to team development but will also foster personal growth (and balance).
- *Reserve your assessments.* Empowerment goes hand in hand with trust. Good tactics include tracking progress in your head and among your officers. However, it would be counterproductive to give the impression that you are standing over your members with a clipboard. They will feel on display and under pressure depending on the tasks or projects. The takeaway is straightforward: Don't give the *perception* of trust; simply trust.
- *Allow people to be their own assessors.* After you empower people and give them the ability to make their own choices within set parameters, work with them to follow their progress. Asking someone how something went and listening to their own assessment of their performance is noticeably different from externally critiquing them from across the desk. People tend to be their own worst critic, and they will strive to do that much better moving forward. Have you ever seen anyone get defensive with themselves? Me neither.

Empowering Those Who Work with You

The other subset involves empowering our peers. We stand shoulder to shoulder with them in emergencies, and we lean on each other through tough times. We can still empower the people around us by keeping the following suggestions in mind:

- *Be true to the cause.* It is exceedingly common to ride coattails in the fire service, especially when dealing with negativity. The low-hanging fruit of our industry is to be negative and to criticize decisions while playing armchair quarterback. However, when there is an opportunity for one of our peers to succeed, we need to stay true to the department and the cause. We have to give honest opinions and fight the urges we may have to run with the herd on tough issues that may spark debates. Empowering your peers doesn't mean agreeing with them; rather, empowering them entails calling intuitive faults you recognize to their attention and letting them process that feedback.

- *Lighten up on your successors.* Throughout this book, a theme has been the constant tug-of-war between our egos, roles, and perceptions. At some point, we need to embrace our successors—the people who will be in our shoes someday. After the acclimation process, new members need to be lifted up through empowering techniques and given opportunities to shine by their peers. Only then will they learn that empowerment is one of the great ways to grow in the department. As with officers, senior members should look for opportunities to let these other members shine and grow.

- *Be aware of talents.* The entire department suffers if a hidden talent remains hidden. Officers may miss the talents that members exhibit to their peers on a daily basis. Spread the word if you see something that is uniquely commendable for a particular individual. For example, a younger member might be outstanding at getting intravenous access in patients. On noticing this ability, you might make a mental note. But what if you alerted your officers that you have witnessed techniques this member used that could help the whole department? Suddenly this new member is giving a class and feeling valuable to the organization. Without your insight and awareness, this member's talent may never have been recognized, preventing an opportunity for the individual and the department to benefit. That's empowerment.

- *Stay positive.* Negativity is palpable. Too many times, we get down on situations or events that have an upside when looked at in a different way. Have you ever worked with someone who always sees the negative side? They are the ones who complain after free pizza is dropped off by a citizen because it doesn't have sausage on it. Empowerment is virtually impossible when the torches are lit and the pitchforks are raised over *everything* that goes on. Stop. Instead of taking the easy way out, find a solution to problems. Negativity does more damage than you might think and becomes toxic when someone constantly complains.

Don't Underestimate Your Team

Our culture requires regimented actions that produce predictable results. Behind the scenes, though, there are several different plotlines to our individual stories that remain alive until the day we retire. As we go through our careers, we have to keep in mind that although we all have individual goals for ourselves, we should also be aware of the people we work with every day for a third of our lives.

Empowering others is a fantastic way to find the balance within the shift because it satisfies our human needs to feel accomplished and like we are part of a team. Our communal instincts make up a portion of who we are. When we suppress each other and ignore the elements that can empower another person, we are cheating the department, the station, the team member, and ultimately ourselves.

As we move up in the hierarchy and assume more control over others, we are saddled with the extraordinary responsibility of nurturing people on a very personal level through rather impersonal methods. Know your people. Trust your people. Respect your people. Empower your people.

Reflective Prompts

1. We all have opportunities to lift each other up and empower those around us. Make a commitment to empowerment by taking progressive steps: Get to know your people, whether they are your peers or subordinates. Most people know each other's marital status and if they have kids. But do you know their hobbies? Interests? Aspirations? Trips they're planning?
2. Aside from personal lives, we need to know what goals our people have for their careers. Do they want to be promoted? Are they just working at the firehouse to supplement their income for another job?
3. Dig deeper if you have an opportunity. How do they feel about their work environment? Do they feel like they have power to do things at the station? Do they feel they are being held back in some way?

15 All Together

The key to work-life balance is to not be afraid to fail at it. You will drop the ball on something; the trick is to know which balls are made of rubber and which are made of glass.

—Nora Roberts

The Three Traits That Make Us . . . Us

Throughout this book, we have twisted and turned through situations that our characters have had to deal with. The endgame of each chapter has been twofold: to pose scenarios that will be relevant to the topic discussed and to strike a nerve here and there that you can relate to.

There have been melodramatic moments that may seem far-fetched, as well as relatively mundane stories that may not seem earth-shattering. Both types of stories matter. The dramatic fireground stories cause us to stop and pause, thinking about what we would perhaps do differently given the situation. And the subtle interpersonal stories can do just as much.

Hopefully, you have started to realize how integral even the smallest interaction is in our development and maintaining balance in the firehouse. Our lives are a bunch of little moments that fuse together to create us. Good or bad, they bind with similar moments and create who we are. Our job is to lift ourselves and each other up to find the balance in everything we do.

Let this be your mantra:

> *The balance is the internal barometer we use to combine intuition, actions, and feelings to find a calm that fills our lives with contentment and meaning.*

We are multidimensional creatures that need to nourish each of three essential traits—namely our mind, body, and spirit. These three traits are no secret, having been the subject matter for numerous authors, teachers, motivators, and coaches. When all of these have a balance to them, our lives become a meaningful and fulfilling journey. If one of them becomes out of balance, we can hopefully use the myriad tools provided in this book to right the ship.

The Mind

Our most powerful tool at our disposal is what usually gets us in trouble most frequently. After all, the nerve center of our body controls everything else. Our mood, intelligence, reasoning, actions, and inactions are all at the mercy of that computer between our ears.

Because this is where our emotions (unfortunately) are controlled, our perception can get distorted. This can lead to overreaction and knee-jerk reactions. To that end, this section details several points to keep in mind (go figure).

Gather the Facts

Throughout this book, some of the characters either fared well or could have fared better if they had gathered all the facts heading into the scenarios they were placed in. Sound decisions and levelheaded conclusions can be reached only if we have everything at our disposal.

In chapter 10, we met Kevin and Trina. Trina was left in the dark because of Kevin's reluctance to talk about a child who had died on a tough emergency medical services call. Once Trina was able to get all the facts about Kevin's internal struggle over the event, she was able to help her husband deal with his emotions. Kevin, like many of us, chose to internalize his problems, and like many of us, he couldn't handle the problem efficiently; consequently, his emotions bled out, materializing in uncharacteristic behavior. Communication is a powerful link to sound judgment. Placing all the cards on the table alleviates a ton of guesswork.

In chapter 4, Chris was a new hire who was on the tip during his first fire. His running mate, Steve, had his back. The problem arose when Steve vacated his task briefly to deal with a problem that arose with the water supply. Chris's mind then started racing, opening the floodgates to all manner of hypotheticals with horrible outcomes—none of which were true. He wasn't given all the facts. Because of his lack of experience, he didn't recognize that Steve never left his support role. The mind is one of our most powerful allies—and our worst enemy.

Accept the Learning Curve

Like most jobs, we are hired into a new position and must deal with the fact that the best teacher is experience. As discussed at different points in the text, we work in an industry where lives are on the line. This can cause enormous amounts of stress and anxiety. No matter what circumstances we find ourselves in, keep in mind that the path we are taking is the only one. Try to make every new experience a teachable one.

Lexi, the main character in chapter 5, was a midlevel firefighter forced into a situation she had little experience with. With her officer not available, she needed to make judgment calls and run a working fire scene that made her uncomfortable. Only through a coordinated effort between her and her peers was she was able to avert disaster until mutual aid officers arrived.

All the characters learned lessons that day. We rise to challenges or sink away from them. Lexi rose to the challenge and realized that things won't go seamlessly all the time. In our mind, we tend to view polarities as the only two possibilities: Either things will go perfectly or they will go horribly. When we think in these extreme terms, we will either get a rush of relief or a constant barrage of ineptitude—all based on how our mind told the story before it even happened.

See the Forest Through the Trees

At the core of our emotional health is the realization of the big picture. What are we here to do? Help each other and be happy. Don't get wrapped up in the little things because they will cumulatively turn into a big one eventually.

Every day presents challenges the moment we get up. There are deadlines to meet, tasks to complete, and problems to solve. Our busy schedules bring with them a number of challenges. How we choose to deal with them is our decision. Find the garden of tranquility in your head and reside there the best you can.

Whatever brings you peace internally, make that the untouchable place in your mind. It becomes the retreat that centers your emotions, quiets your thoughts, and is a voice of reason that negates the swirling negativity that inundates us daily.

Look at the big picture of your life occasionally and take stock of where you are. You'll be surprised at just how good you have it. If your overall perception of your life is damaged and sad, then address it. If it is enjoyable and brings a smile to your face, then don't let small things permeate the bubble you've created. Your mind, thoughts, and life are worth so much more than the energy spent on negatively reacting to your problems. There are always other ways to get through the tough times in your life.

Everyone's World Does Not Revolve Around You

We look at ourselves through a powerful microscope. Moreover, we tend to look at ourselves in terms of how we think other people see us. However, most of the time, they don't see us the way we think they do. Stop wasting precious energy on trying to figure out what others think of you.

An exceptionally good tactic you can use to ensure that the people around you have exactly the opinion you want them to have is to stay true to your principles and be honest with yourself. People may not always agree with you and your opinions. But you will receive a level of respect from people because you are authentic and genuine.

Almost everyone can see through a facade. Don't fake your way through life. Express what you know, ask questions about what you don't, and show enough humility to know the difference between the two. We're all imperfect. It takes massive amounts of energy to be perfect, and it takes twice that amount of energy to act as if you are on top of it. You be you.

Stay Neutral

One strategy you can use to calm your mind is to never get too high or too low. It is never as good or as bad as we think, and by staying neutral we can swing the needle where we need to in stressful situations.

In chapter 3, Ray was highly regarded as a legend on the department. He had been through the wars and had earned enough accolades to cement his legacy without doing anything else the rest of his career. He chose to stay neutral when he could have let his head swell.

When Ray was confronted with the situation between the midlevel firefighters and a new hire, he arrived at a choice: maintain his neutrality or go full steam ahead with a less than heroic decision. He chose to maintain his neutrality, and the results were positive.

Staying neutral does not mean you are devoid of any emotion at all. It simply means that you don't dwell on good or bad fortune from occurrence to occurrence. Celebrate the good, mourn the bad, and move on. Everything is fleeting in our lives. Keeping neutral allows us to be at our peak level of awareness so we are not surprised at what may lurk around the next bend.

The Body

The physical demand placed on us in the fire service has been well documented—with good reason. We get the stuffing beaten out of our bodies. The physical toll the job takes on us wears us down over time, and if we don't keep our house in order physically, then we suffer over the long haul. This section details a few ways to keep our physical well-being in balance.

Engines Run Better on Premium Fuel

This isn't a book on nutrition. Still, it is worth mentioning that when we put garbage in our bodies, our bodies run like garbage. I'm not talking about eating cookies and ice cream instead of fruits and vegetables. The *vast* majority of members I know eat well and watch their diets while on shift, yet let it rip when it comes to drinking alcohol and completely gorging themselves when they aren't on duty.

That leaves our bodies with a whole lot of sorting out to function right. One phrase really encapsulates the rule of thumb for eating and drinking: "Everything in moderation." If you follow this maxim, you will usually give your body enough time to process all of what we enjoy, instead of calling on our organs to make up for our binging.

Sleep Is Essential

Usually, the mind controls the body. In the natural progression, our minds make decisions that put our bodies to work. However, in one situation, the road actually travels in the opposite direction: sleep. This is the one area where the relationship inverts and the body actually dictates over the mind.

When we lack good rest, the mind becomes clouded, and our decision-making abilities become dull and slow. The only thing that counters a lack of sleep is adrenaline. We've all felt what it's like to be exhausted at 3 a.m., but when the tone sounds, we are out the door and in our bunkers with the blood flowing and our minds racing. That's adrenaline taking over. However, that is a quick fix—and a temporary one. True, we will put out the fire and perform what we have to because someone's life could be on the line. But eventually we come down from the high we were on.

The same holds true for emergency medical services calls. When someone's life hangs on our next decision, we push the right drug and perform the appropriate intervention. Pay attention to how fast you get tired when the emergency is over, though. Whether you are packing hose after a structure fire or cleaning up the squad after a full arrest, your body will deplete that adrenaline, and you will become exhausted.

Not getting enough sleep is something most of us deal with in the service. Years tick by. Soon we've been on the job for 2 decades with scattered naps being our only true rest throughout our careers. It takes a toll. Get enough rest. Your body will thank you for it.

Change Routines for Your Changing Body

Our bodies change as we get older. There is no way we can lift as much at 50 as we could at 20 years old. Don't try to be the exception to that rule. That's how muscles get pulled, shoulders tear, and knees need to be replaced.

As we get older, we need to turn to other ways of staying in shape than the heavy lifting of our youth. For example, consider doing more pulley work in the gym. Stop holding onto the old routines.

Sooner or later, something is going to give out, and all the sick time you've hoarded for a payday at retirement will evaporate in the matter of a few months. Was it worth lifting all that weight at the gym when nobody really cared how much you bench-pressed? I'll answer that for you: Nope.

The Spirit

At the center of our discussion lies the third trait we possess: the spirit. We are all on a spiritual journey. Some of us are lucky enough to recognize the ride we are on through careful self-reflection and good overall awareness. However, most of us have trouble connecting to our spirits owing to the noise we have surrounded ourselves with. This section outlines a few considerations to keep in mind as you pursue the balance in your spiritual journey.

You Grow When You Let Others Grow

Letting others grow and allowing them to be who they truly are capable of seems so simple. Yet it is one of the hardest things you will ever do in your life, let alone your career.

Because of the mental gymnastics we constantly do in our heads, we tend to hold people back, either deliberately or by accident. Sam, the captain in charge of the scene in chapter 8, fought all his instincts to jump into the action and take over the hands-on duties from his subordinates. He was used to acting and instead had to (rightfully) watch others do those actions. He ultimately succeeded, but it was an uphill battle. He strengthened himself by strengthening others.

Relinquishing control over others is incredibly liberating if you can do it. Which leads to the next point . . .

Stop Trying to Control Others

In chapter 14, Sue was being held back by her lieutenant, Cole. It wasn't intentional, but by trying to be thorough in his duties, Cole was constantly keeping tabs on his members. The result was the stunting of one of his member's personal growth. Sue was frustrated and couldn't really put her finger on the reason why she felt so unfulfilled until her old boss was able to put the pieces together and solve the mystery.

Guiding your people is completely different from controlling them. Setting parameters and expectations allows them to bounce around and do what they need to. Suffocating them by controlling their every movement is a power play that will be self-limiting.

As a crude analogy, ask yourself which one your dog probably prefers, having an invisible fence with a huge backyard or being on a 2 ft leash with a choke collar? We both know which it would prefer.

Control is a lethal word, tightly linked with insecurity and doubt. Don't let your insecurities dictate your relationship with the people you work with. It usually doesn't end well.

Keep Your Ego Locked Up

Chapter 3 also introduced Bryan and Cody, two midlevel firefighters whose insecurity led them to take out their fears on the young new hire before dinner. Similar situations unfold in real life more than we realize. If you pay close attention, in many firehouses, flare-ups are occurring all the time that make it clear that some members aren't in control as much as they believe.

Smitty and Bud in chapter 13 used self-inflation to control what others thought of them. This tactic is much less damaging than a direct attack on another member. Although it seems harmless enough, it is still an attempt by two members to create a perception instead of letting the perception come to fruition on its own.

In chapter 6, Chuck, a senior member of his department, beautifully encapsulates keeping his ego under control. Only when he had to assert himself did he throw his experience and discipline at younger members in harm's way. He had balance throughout the story, and he was *still* in balance once he made the decisive intervention on the fireground that may have saved lives.

Always keep an eye on your ego and make sure that it comes out to play only when necessary—which is never.

Remember, There's Always a Way Through

In chapter 11, we met Jake. He had a drinking problem that spiraled out of control to the point where he needed to be pulled off the fireground because of the potential danger he posed to his peers. Jake turned to alcohol without recognizing that it was his way of coping with the stress he had endured for a long time in his adult life.

It isn't always easy to make the right choice. Sometimes the right path isn't necessarily the easiest. But when we take an honest look at ourselves and stop making excuses, we can assess whether we are on the right path toward our balance.

Many times, when we make poor choices to deal with adverse events or circumstances, it is viewed—seemingly innocuously—as decompressing or taking a break. Whatever you want to call it, if you are compelled to turn to some type of abuse to deal with your situation, it is a problem. This has a direct effect on your spirit, believe it or not. Instead of keeping your ego and urges under control, you are suppressing your true nature instead.

You may be able to sustain yourself for a while this way, but sooner or later, it will catch up to you. The sooner you face your demons, however small, the sooner you can move to being happy and fulfilled in your life.

It's a Wrap

I hope this book has made you stop and think about where you are in your growth as a firefighter. We are all growing together in this crazy profession. Let's lean on and learn from each other.

Finding the balance is a constant activity that is worth the work and self-reflection if we put in the time and effort to take our powerful perception and direct it inward. Be honest with yourself when you take inventory of where you are.

The balance is the secret ingredient we can all attain. Help yourself and you will help those around you. Balancing yourself out gives us a great opportunity to pass it on, and it demonstrates what is possible to others without even trying.

Be happy and realize that we work in one of the greatest professions ever to grace humanity. Take pride in your work. Love your family. Respect your coworkers. And remember, at the end of the day, at the end of our careers, at the end of our lives, we are all in this together.

Reflective Prompts

1. What are the proudest moments you've had as a firefighter? What are the calls that have been lodged in your memory as your finest moments?

2. What were the saddest days of your professional life? Have you learned from those and grown from them? Or do they still linger with you after years of thought and processing in your mind?
3. If you are just starting out in your career, was there a particular event that inspired you to choose this path in your life?
4. Is there any facet of your firefighting journey where you feel you are "settling" for circumstances? What steps can you take to change matters and inject energy back into your career?

Index

A
academy training 26–29
accepting change 53–55
accountability
    leadership and 65
    motivation and 3
    self 45
acting versus directing 77
addiction
    addressing 117–118
    balance and 11, 157
    drinking and 2–3, 7–8, 159–160
    qualities of 118
    resources for 120
    stress and 11, 110–114, 116–117
administrative challenges 2, 67
administrative tasks and chiefs 90
administrative transparency 92
AFD (Avon (OH) Fire Department) 110
aging and firefighting 48–50, 52–56, 158
Ambrose, John xi, xvii
anxiety
    home 102
    leadership 52–53, 72–73
    on-site 29–30
    rookies and 26, 154
attitude. *See also* negative attitude
    adjusting one's 126
    empowerment and 150–151
    personal responsibility and 22
    positive 150
authenticity
    negativity and 21
    professional roles and 22
    respect and 21, 156
authority
    balancing 78–79
    challenges to 78
    exercising 41

    organization of 90
    positive use of 19–20
    responsibility and 61–62
Avon (OH) Fire Department (AFD) xi
awareness
    ego and 139–140
    emotional 106
    empathy and 106

B
balance
    addiction and 11, 157
    authority and 78–79
    comfort vs. 4
    defining 3–4
    empowerment and 147–151
    family life and 2, 7, 33, 137
    leadership and 65, 90–91, 146–147
    roles and 20, 77
    self-reflection and 4
    stress and 103–104, 137–138
    work-life 5, 101–106
    workplace behavior and 20, 54, 64
    workplace culture and 54, 78–79
    workplace relationships and 60–63
banter 20
behavioral change 18, 112–114
behavioral solutions 19
behavior and reprimanding 62
belonging 52–53
boundaries in relationships 31–32
budgeting 8, 32

C
captains
    advocacy and 80
    decision making as 79–80
    Meadows, Rodney 32
    new 71–82
    workplace culture and 78–80

career benchmarks 22
career satisfaction 2, 144
central fire stations 128
chain of command
    transparency and 65
    two-way 89
    utility of 93
change
    accepting 53–55
    behavioral 18
    professional growth and 59
    professional roles and 77
    senior firefighters and 53
    workplace culture and 48–50, 54
chiefs
    administrative tasks and 90
    communication and 89–92
    connecting with members as 92
    Croker, Edward 10–11
    decision making as 89
    humility in 93
    responsibilities of 90
    suburban vs. metropolitan 90
    transparency and 91–92, 94
clarity in management 148–149
collaboration and experience 68
comfort vs. balance 4
command presence 21, 77
common interests and relationships 57
communication
    chiefs and 89–92
    emotions and 64, 100, 104, 154
    family 33–34, 43, 98–101
    family and 103
    fire departments and 43–44
    leadership and 21, 92
    relationships and 30–31, 43–44, 130–131
    stress and 102, 104, 119
    successful 30, 39–40
compartmentalizing emotions 9, 101, 104–105
complacency in leadership 81–82
confidence
    firefighters and 26
    promotions and 66
    rookies and 26
    training and 40–41
conflict management 78–79, 91, 94–95
constructive criticism 41, 80, 146–147
control and management 159–160
control and relationships 159

coping mechanisms 11, 119–120
costs of living 2
coworkers
    age gaps and 52–56
    extended time with 10
    promotions and 66
    responsibility and 13
    supporting 19
creativity and empowerment 148
Croker, Edward (chief) 10–11
culture in fire departments 33, 54–57, 125

D
death 12, 100
decision making
    captains and 79–80
    chiefs and 89
    effective 38–40
    patience and 51
    quick 10, 30
    sleep and 157
Deferred Retirement Option Plan (DROP) 2, 47
delegating 89
department advocacy 92
dependents 3
dialectic traits in firefighters 10
divorce 3
drinking
    addiction and 2–3, 7–8, 159–160
    behavioral change and 112–114
    stress and 116–118
DROP (Deferred Retirement Option Plan) 2, 47

E
ego
    awareness and 139–140
    empathy and 140–141
    family and 138
    fear and 138
    grief and 136–137
    managing 10, 12, 139–141, 159
    relationships and 137–138
    senior firefighters and 136–138
emotions
    anticipating 12
    awareness and 106
    communication and 64, 100, 104, 154
    compartmentalizing 9, 101, 104–105

delaying 9
family and 97–100
grounding 155–156
lack of 105
managing difficult 101, 154
responding to 63–64
rookies and 154
transparency and 102–103
empathy
awareness and 106
ego and 140–141
leadership and 81
employee assistance programs 11, 114
empowerment
attitude and 150–151
balance through 147–151
creativity and 148
individuality and 150
leadership and 147–151
management and 147–151
professional outcomes of 147–148
professional ranks and 151
succession and 150
teamwork and 147
trust and 149
energy levels 22
experience
collaboration and 68
lack of 130
learning from 155
routine and 29

F

family
balancing 2, 7, 32–33, 63–65, 137
communication and 33–34, 43, 98–101, 103
ego and 138
emotions and 97–100
leaning on 42–43
relationship to 7, 103
transitions in 55–56
fear and ego 138
fire departments
Avon, OH xi
central 128
communication and 43–44
culture in 125
lineage and 55, 125–126
metropolitan 90–91, 125–127
officers and 124
rural 129–132

slow 35–44
suburban 127–129
tactical approaches in 126–127
types of 125–126
volunteer 36, 89–90, 122–132
firefighters. *See also* rookie firefighters; senior firefighters
confidence in 26
dialectic traits in 10
middle-tier 41–42
responsibility and 4
volunteer 36, 89–90, 122–132
firefighting
aging and 2, 48–50, 52–56, 158
gender bias and 11
heroism in 19
lineage in 9, 52, 55, 125–126
physical expenditure and 10, 52–53, 156–158
public reputation in 12
social aspects of 31–32
stereotypes in 48
three levels of 45

G

gender bias 11
goals
achieving 5
chasing 11
leadership 80–81
realistic 90
self-serving 21
gratitude 155
grief and ego 136–137

H

hands-on learning 57
health
nutrition and 157
routine and 42
sleep 157
herd mentality 21
heroism 19
holidays 7, 12
home anxiety 102
humility in chiefs 93

I

imbalance 4
improvement in leadership 70
individuality and empowerment 150
ingenuity 130

insecurity 21
instant gratification 117–118
intuition 56

L

leadership
  accountability and 65
  anxiety and 52–53, 72–73
  balance in 65, 90–91, 146–147
  challenges in 77, 91–92
  communication and 21, 92
  complacency in 81–82
  conflict management and 78–79
  delegating and 89
  empathy and 81
  empowerment through 147–151
  goals in 80–81
  improving in 70
  teamwork and 68
  transparency and 91–92, 94
  trust and 65, 69–70, 80
lineage in firefighting 9, 52, 55, 125–126
listening skills 68

M

management. *See also* conflict management
  clarity in 148–149
  control in 159–160
  ego 10, 12, 139–141, 159
  empowerment and 147–151
  hierarchy of 67, 93
  micro 89, 146–147
  officers and 44, 60–61
  relationships and 79–80
  styles of 146–147
mayday scenario 77
Meadows, Rodney (captain) 32
mentorship
  exemplary 52–53
  professional roles and 22–24
  senior 22, 31, 52–53
metropolitan departments
  chiefs in 90–91
  culture in 125
  integrating into 126–127
middle-tier firefighters 41–42
motivation
  accountability and 3
  evaluating 11
  ulterior 21

N

negative attitudes
  authenticity and 21
  detriments of 149–150
  role models and 23
new experiences and rookies 25–29
9/11 xvii
nutrition and self-care 157

O

officers
  department size and 124
  role of xi
  team management and 44, 60–61
outreach programs 11

P

passion 2, 34, 42
patience 51, 67
perfectionism 41, 147–148
personal responsibility and attitude 22
physical ability and age 48–50, 55, 158
physicality 10, 48–50, 52–53, 156–158
positive attitude 150
preparedness 25, 45
pride 11, 93, 136–137
process people 9
professional growth
  change and 59
  patience and 67
  reevaluating 42
  stages of 93
professional ranks
  behavior differences in 18
  climbing 10
  empowerment through 151
  respect and 147
  responsibility and 19–20, 41, 55–56
professional reputation 18
professional roles
  authenticity in 22
  change in 77
  insecurity and 21
  mentorship through 22–24
  promotions and 22
  relationships and 20–24, 65
  tenure and 18
  types of 22–23
programs
  addiction recovery 120
  employee assistance 11
  retirement 2, 47

promotions
    addressing 66, 91
    challenges of 65–68
    confidence and 66
    coworkers and 66
    professional roles and 22
public reputation 12

R

relationships. *See also* workplace relationships
    boundaries in 31–32
    common interests and 57
    communication and 30–31, 43–44, 130–131
    consistency in 106
    control and 159
    ego and 137–138
    family 7, 103
    roles and 20–24, 65
reprimanding poor behavior 62
respect and authenticity 21, 147, 156
responsibility
    authority and 61–62
    chiefs and 90
    coworkers and 13
    firefighters and 4
    professional rank and 19–20, 41, 55–56
    roles and 22–23
retirement 2, 47
role models 23–24
roles. *See also* professional roles
    balancing various 20, 77
    determining your 22–23
    officer xi
    relationships and 20–24, 65
    responsibility and 22–23
    three categories of 22–23
    utility of 22
rookie firefighters
    academy training and 26–29
    anxiety and 26, 154
    confidence in 26
    emotions and 154
    new experiences and 25–29
    treatment of 15–20
routine
    establishing 29
    experience and 29
    healthy 42
    prioritizing 33
rural departments 129–132

S

sarcasm 43, 49
self-care 105–106, 157
self-reflection 4
self-worth 22
senior firefighters
    change and 53
    ego and 136–138
    intuition and 56
    mentorship and 22, 31, 52–53
    respecting 52–55
    stagnancy and 53
    workplace culture and 54
sleep 157
slow fire departments 35–44
socialization 31–32
social skills and stress 103
stagnancy 53
State of the Department address 92
stereotypes 48
storytelling 133–137
strength of character 34
stress
    addiction and 11, 110–114, 116–117
    adverse reactions to 114, 117
    balance and 103–104, 137–138
    communication and 102, 104, 119
    drinking and 116–118
    effects of 11–12, 103
    five senses and 115–116
    long-term 116–117
    on-site 26
    responding to 11, 102, 119
    signs of 103–104
    social skills and 103
    triggers of 64, 114–116
structure collapse 123–124
subordinates
    advocating for 69
    listening to 68
    transparency and 91–93
suburban fire departments 127–129
succession and empowerment 150
switching departments 125

T

tactical approaches 126–127
teamwork
    effective 50–52
    empowerment and 147
    importance of 40–41
    leadership and 68
technical difficulties 27–28

tenure 18
time off 7
tolerance 57
tradition 18, 52–53
training
    confidence and 40–41
    rookie 26–29
    types of departments and 128
transparency
    administrative 92
    chain of command and 65
    chiefs and 91–92, 94
    emotional 102–103
    leadership roles and 91–92, 94
trauma 101, 119
trust
    empowerment and 149
    leadership and 65, 69–70, 80
    workplace relationships and 129

U

ulterior motives 21
understaffing 40–41

V

volunteer departments 36, 89–90, 122–132

W

work environments 10, 25
work-life balance 5, 101–106
workplace culture
    balancing 54, 78–79
    captains and 78–80
    changes in 48–50, 54
    establishing 67
    integrating into 33
    senior firefighters and 54
workplace relationships
    balancing 60–63
    boundaries in 31–32
    common interests and 57
    control in 159
    ego and 137–138
    employer and employee 79–80
    suburban departments and 128
    tight-knit 128
    trust and 129
    types of 31